BIG CATS

Edited by **Avadh B. Shrivastav**
and **Keshav P. Singh**

Big Cats
http://dx.doi.org/10.5772/66582
Edited by Avadh B. Shrivastav and Keshav P. Singh

Contributors

Hussein Abd-Elhay Kaoud, Lucas Gonçalves, Edward Narayan, Nagarajan Baskaran, Janice Vaz, Juan Jose Alava, Miguel Saavedra Mendoza, Paul Cun, Eric Horstman, Sonia Carabajo, Santiago Jimenez Gonzalez, Claudia Jimenez Escobar, Manuel Ruiz-Garcia, Jesus Maldonado, Haydy Monsalve, Ligia M. Jimenez Robayo, Susan L Castro, Jimmy J. Vargas, Sonia R. Quintanilla, Alberto Sanchez, Daniel Veraguas, Diana Echeverry, Fidel Ovidio Castro, Lleretny Rodriguez-Alvarez, A. B. Shrivastav, K.P. Singh

Published by InTech
Janeza Trdine 9, 51000 Rijeka, Croatia

Publishing Process Manager Maja Bozicevic
Technical Editor SPi Global
Cover InTech Design team

First published December, 2017
Printed in Croatia
Legal deposit, Croatia: National and University Library in Zagreb

Additional hard copies can be obtained from orders@intechopen.com

Big Cats, Edited by Avadh B. Shrivastav and Keshav P. Singh
p. cm.
Print ISBN 978-953-51-3683-5
Online ISBN 978-953-51-3684-2

For EU product safety concerns:
IN TECH d.o.o., Prolaz Marije Krucifikse Kozulić 3, 51000 Rijeka, Croatia,
info@intechopen.com or visit our website at intechopen.com.

Meet the editors

Dr. Avadh B. Shrivastav, recipient of Saheed Amrita Devi Vishnoi Award for wildlife conservation, is a coordinator for health management practices in Central India and a former Director of School of Wildlife Forensic and Health in NDVSU, Jabalpur, still providing his expertise . Professor Shrivastav is a veterinary pathologist who trained at Wildlife Institutes of India and National Wildlife Health Center, USA. He was a principal investigator of 11 research projects on wildlife health including 2 prestigious research projects, namely "Niche Area of Excellence on Wildlife Forensic and Health" (ICAR) and "Centre for Wildlife Forensic and Health"(State forest Department) funded by ICAR, New Delhi, and Government of Madhya Pradesh, respectively. Wildlife Institutes of India, Central Zoo Authority of India, and World Wildlife Fund for Nature are availing his expert services. He has published 140 research articles in the journals of repute and guided 55 veterinary students for Master's and PhD degrees.

Keshav P. Singh is working as a Wildlife Biologist at the School of Wildlife Forensic and Health, Nanaji Deshmukh Veterinary Science University, Jabalpur, MP, India. He earned his PhD degree in the field of Veterinary Protozoology and contributed in establishing haemato biochemical profile of wild animals subjected to parasitic diseases. A professional skill of Dr. Singh is substantiating the novelty in research, teaching and extension activities in the field of wildlife health management for conservation of biodiversity in protected and non-protected forests. During his entire academic career, Dr. Singh has published a "Booklet on Health Management of Wild Animals" field guide, book chapters and review articles including both research and popular articles in the national and international journals of repute in addition to supervising Master's and PhD degree thesis research work of veterinary students on wildlife health management aspects.

Contents

Preface

Wildlife conservation in the developing world is an important and essential component of healthy ecosystem, where human being is also one of the living creatures. Wildlife conservation means to protect survival of the species in their natural ecosystem. Unfortunately, due to degradation in the quality and quantity of habitat, it becomes a critical factor for the survival of many species, particularly big cats, and brings them under the flag of endangered category in different geographical areas of the world.

The scientific contributions in the book are from authors from diverse disciplines who are also actively involved in wildlife ecology, health management and genetic studies of the majestic species. The texts incorporated in the book are an attempt to provide a comprehensive knowledge on free-range and captive felines pertaining to their global distribution, taxonomic status, health management, and conservation physiology. The authors of each chapter have given consolidated information on ground realities. The book will be useful to students of biological sciences, veterinarians, researchers, and park managers.

The editors are thankful to every individual who helped in the preparation of this book. The book has future aspects with innovative tools for conservation of big cats in different protected and non-protected forests of tropical and temperate countries.

The editors are also thankful to Prof. P.D. Juyal, Hon'ble Vice-Chancellor, Nanaji Deshmukh Veterinary Science University, Jabalpur, India, for giving them permission to edit the book. The editors are also indebted to the chapter contributors for accepting helpful criticism for the present shape of the book.

Last but not least, thanks are due to Ms. Maja Bozicevic, Publishing Process Manager of InTechOpen, for sending information and guidelines for editing the book chapters well on time.

Dr. Avadh B. Shrivastav and Dr. Keshav P. Singh
School of Wildlife Forensic and Health, NDVSU
Jabalpur, MP, India

Section 1

Health Managment

Chapter 1

Introductory Chapter: Health Management of Big Cats

Avadh Bihari Shrivastav and Keshav Pratap Singh

Additional information is available at the end of the chapter

http://dx.doi.org/10.5772/intechopen.72049

1. Introduction

Nature has imparted resources for survival and sustenance of life on the Earth. About 8.7 million (6.5 million on land and 2.2 million under ocean) known species of flora and fauna are dispersed in the different geographic zones, and some of them are indicator species. The big cats have been playing a vital role in establishing population equilibrium among primary, secondary and tertiary consumers in the sylvatic food chains [1]. The big cats had been placed taxonomically under Felidae with four subfamilies, viz. Pantherinae, Felinae, Machairodontinae (extinct) and Proailurinae (extinct). However, in the present status, there are only three regions around the world where population of felines exists, viz. Africa (cheetah, leopard and lions), Asia (Asiatic cheetah, Amur tiger, clouded leopard, Sunda clouded leopard, snow leopards, Asiatic lions and tigers) and North and South America (cougars and jaguars; Spanish lion and tigers). According to the geographical background of *Panthera* species, tiger, lion and cheetah are evolved from their prehistoric creature called cave lion. Despite of infighting and retaliate killing, the disease manifestations in big cats were underestimated albeit emerged as serious threats for their survival.

In the present scenario, the complexities of challenges are gradually expanding through encroachment of forest land and increasing interaction between humans and domestic and wild animals leading menace to wildlife. Nonetheless, the admixture process of companion animals in the free-ranging habitat for sharing grasslands or water holes is a key factor for disease transmission in either of wild or domestic animals [2]. Indiscriminate uses of forest wealth and simplification of the ecosystem are determined causes for unwarranted migration of zoonotic pathogens from sylvatic to domestic cycle and vice versa [3]. Many pathogens of bacterial, viral as well as parasitic origins have been changing their migration pattern owing to consistent human interventions in the forest habitat. Therefore, susceptibility of zoonotic diseases is also increasing and expanding their host ranges day by day. The tribal population of adjoining areas of national parks and sanctuaries is more prone to such zoonotic infections as they utilise the common water resources.

It has been believed that free-ranging wild animals are comparatively possessing a high standards of health which is rigidly maintained by the action of natural selection. But owing to increased human interference, poaching and destruction of habitat, most of the significant threats are health related [4]. Therefore, the use of advance technology for health monitoring, disease diagnosis and forensic investigations and their legal use are need of the hours for conservation strategies of wild animals.

2. Status of health management practices

Till the recent past, emphasis on ecological management has been focused on strengthening and provoking natural devices for conservation of wildlife wealth, whereas health management aspects were out of focus. Only firefight approaches were made out, and it was sometimes exclusively restricted to rescue of ailing wild animal. Up to 1990, there were limited amenities for restraining of wild animals, whereas after introduction of chemical immobilisation and modern technologies, the task became far from fear and turned into a mandatory component for drug orientation and health monitoring of felines [5]. Large-scale mortality in the African lions in Serengeti Reserves in South Africa due to canine distemper and also due to *Feline panleukopenia virus* in 1995–1996 in different parts of the world including Japan and India was the eye opener in an Indian subcontinent. Subsequently, the government and national and international NGOs forced to think and adapt scientific wildlife health management strategies in their conservation programmes.

The present book has been focused on big cats and probably would be a tool for conservationist, wildlife biologists and wildlife veterinarians for making a concrete research programme to study anatomical variations, habit, habitat and their health aspects, including treatment, disease diagnosis and also active veterinary interventions, for their sustenance and survival in the free ranging as well as captive conditions.

The following are the key points for the successful health management programme either of captivity or in free-ranging habitat:

I. **Care and management in zoos/captivity:** The main objectives of zoos and captivity are conservation education, breeding of critically endangered species, recreational values and also scientific studies on various aspects related to behaviour and health of the animal species. However, the risk factor during health monitoring and treatment is always around the veterinarians, but following the standard operating procedures and the use of diagnostic tools, the handling of situations both in free ranging and captivity may ease the task. The handling of wild animals in captivity is quite complex and challenging, which needs knowledge about habit and health status of particular species. Sometimes, they have camouflage attitude showing false sickness, aggressive behaviour or even unpredictable posture during collection of biological samples for disease diagnosis [6]. Therefore, during health monitoring or shifting of animal from the zoo, the imitation of standard guidelines shall be followed (**Figure 1**).

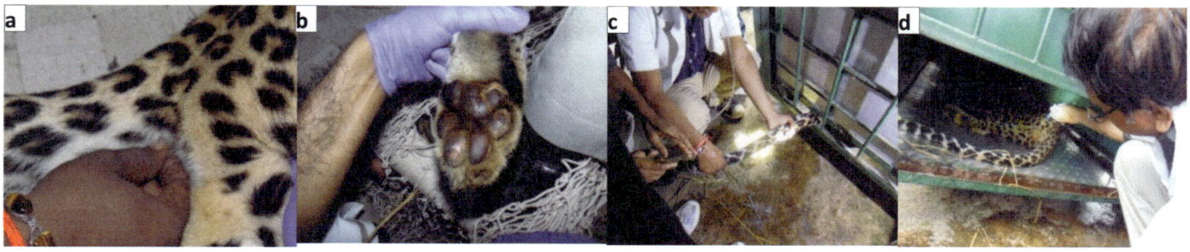

Figure 1. Health monitoring of ailing captive leopard: (a and b) monitoring of skin coat and cushion pad injury; (c) using caudal vein for blood collection and drip administration (d) drug administration by intramuscular root.

II. **Analysis of health aspects in free-ranging habitat:** The assessment of disease manifestations has yet not been fully warranted, but veterinary scientists have developed a predictive model through clinical symptoms, changes in behaviour as well as their physical appearances. The ailing wild animal subject to assess the risk of diseases and needs to be attentive for their care well in time. In free-ranging animals, veterinary interventions are difficult and need multidisciplinary approach for restraining, radio collaring, treatment or collection of biological samples for laboratory investigations to evaluate health status (**Figures 2** and **3**).

III. **Precautions during immobilisation:** The most adventurous and mandatory tasks of veterinarians are confined to restraining, radio collaring, health monitoring, disease diagnosis and reintroduction of wild animals. In such conditions, calculation of doses and combination of drugs for each and every circumstance seem rather difficult, and only a trained and experienced wildlife veterinarian knows better for how and why requisite doses may be administered to down the animal. The expert opinion/guidance may be helpful in commencing the plan; permission from the competent authorities is essential. The operational team must be well equipped along with sufficient number of wildlife veterinarians for successful operation and also to minimize the chance of errors [4]:

   i. Every step, in handling and in also administration of drugs in big cats, must be introduced slowly and carefully to avoid flight fear reaction and stress.

   ii. The use of appropriate bait prior to immobilisation/restraining of targeted animal.

Figure 2. Health monitoring and radio collaring of tiger: (a) immobilised tiger; (b) collection of blood, urine and faecal samples; (c and d) radio collaring for ecological studies and revived tiger.

iii. For captive animals, the physical presence of the caretaker is a must during handling or restraining.

iv. Blindfolds must be used; excessive noise and touching stress should be minimised.

v. Excessive and rough handling or stress in the restrained animal can lead to hyperthermia and capture myopathy particularly during warm or hot climate; therefore, such incidents may be avoided.

vi. Capturing and restraining of pregnant animals may be avoided.

vii. Physical posture of the sedated big cat must be in normal sitting posture to ensure that animal's smooth breathing should not be compromised.

viii. It is important that handler should protect themselves against possible injuries, exposure to drugs or chemicals, animal excretions, etc. to avoid zoonotic infections.

IV. Study of emerging disease threats: The infectious diseases like *Feline panleukopenia virus*, canine distemper, feline viral rhinotracheitis, feline calcivirus infection, feline infectious peritonitis, feline immunodeficiency virus, ehrlichiosis, trypanosomiasis, babesiosis, paragonimiasis and gnathostomiasis are fatal; therefore, their prevention and control are important both in captivity and free-ranging populations. On the other hand, the laboratory diagnostics are important; thus, appropriate samples in proper preservative are important for diagnosis.

V. Post-mortem examination: The proverb has been very common among pathologists as 'the carcass never tell lies providing expert knows the language of carcass'. Really, it means as the correlated post-mortem changes may lead to the past history and situation by interpreting the lesions present on the external surfawce and internal organs (**Figure 4**). On the basis of the presence of lesions, tentative diagnosis may be ascertained and communicated to competent authority for appropriate action.

VI. Study of human and wild animal interface: Human and domestic animal intermixing with wild animals may communicate to zoonotic disease particularly tuberculosis that poses a threat to the health of wild animals. It is also responsible for human wildlife conflict resulting into poaching, poisoning and electrocution of problematic or distressed wild animals. So there is a need for disease surveillance plan along with scientific views to overcome human wildlife conflicts.

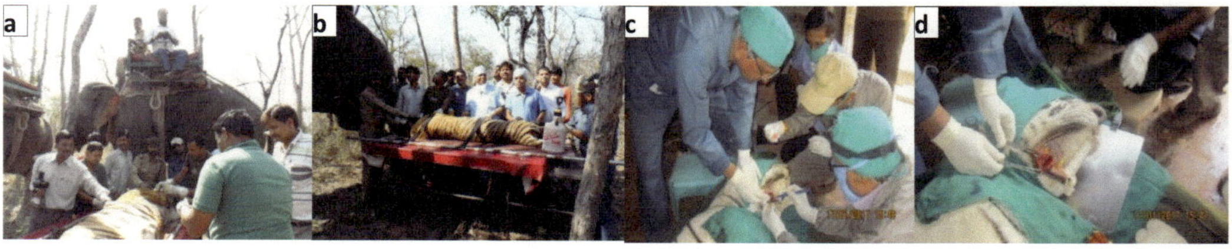

Figure 3. Surgical interventions for severely injured free-ranging tiger (a and b) and for extraction of neoplastic growth in oral cavity of a captive tiger (c and d).

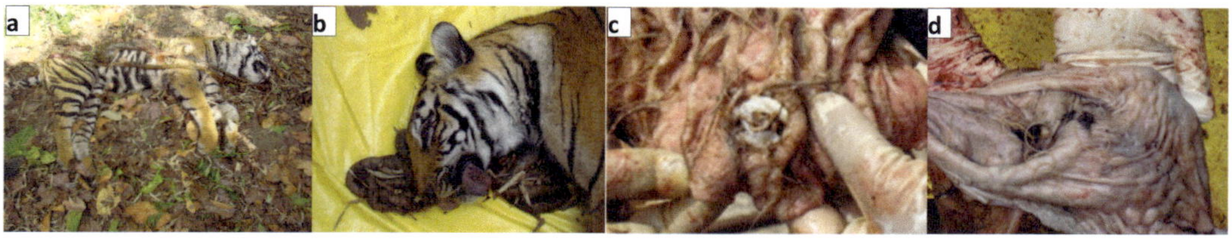

Figure 4. Post-mortem examination of carcass of a tiger. (a) Emaciated carcass. (b) Cynotic tongue. (c) Nodular worm (*Gnathostoma spinigerum*) in the stomach. (d) Presence of *Toxascaris leonina* worms and haemorrhagic enteritis in small intestine.

Acknowledgements

The authors are thankful to Hon'ble Vice-Chancellor, Nanaji Deshmukh, Veterinary Science University, Jabalpur, for the permission and moral support to act as editors of the book and Principal Chief Conservator of Forests (Wildlife), Government of M.P. for their kind support in scientific wildlife health management and recognition of the School of Wildlife Forensic and Health.

Author details

Avadh Bihari Shrivastav[1]* and Keshav Pratap Singh[2]

*Address all correspondence to: drabshrivastav@gmail.com

1 SWFH, Jabalpur, India

2 School of Wildlife Forensic and Health, Nanaji Deshmukh Veterinary Science University, Jabalpur, India

References

[1] Gopal R. Fundamentals of Wildlife Management. 1st ed. Dehradun: Natraj Publishers; 2004. pp. 59-90

[2] Artois M, Blancou J, Dupeyroux O, Gilot-Formont E. Sustainable control of zoonotic pathogens in wildlife: How to be fair to wildlife? Revue scientifique et technique (International Office of Epizootics). 2011;**30**(3):733-743

[3] Shah HL. An integrated approach to the study of zoonosis. Journal of Veterinary Parasitology. 1987;**1**(1&2):7-12

[4] Shrivastav AB. Wildlife health: A new discipline: Essential for tiger conservation programme. INTAS Polivet. 2001;**2**:134-136

[5] Shrivastav AB, Singh KP. Tiger Blood: Haematology and Biochemical Studies in the Blood Cell. Croatia: InTech; 2012. pp. 229-242

[6] Miller RE, Fowler M. Fowler's zoo and wild animal medicine. In: Disease Risk Analysis in Wildlife Health Field Studies: Sharon Deem. 7th ed. St. Louis Missouri: Elsevier Inc. Saunders Publication; 2012. pp. 2-7

Chapter 2

Heart Conditions in Felidae

Hussein Abdelhay Kaoud

Additional information is available at the end of the chapter

http://dx.doi.org/10.5772/intechopen.69781

Abstract

In this chapter "Heart Conditions in Felidae," we addressed some facts of the cardiovascular system and its disorders. The disorders are: arrhythmogenic right ventricular cardiomyopathy, atrial fibrillation, atrioventricular heart block, cardiac tumors, chronic valve disease (hypertrophic cardiomyopathy), dilated cardiomyopathy, pulmonary hypertension, pulmonic stenosis, subaortic stenosis, and spongy myocardium. Veterinary clinical evaluation of the feline with heart disease is based on many aspects, and also the deviation from the normal standards suggests but does not specify structural heart disease.

Keywords: Felidae, cardiovascular system, cardiac disorders, blood parameters, therapy

1. Introduction

Feline or cats (family Felidae) consist of 37 species of cats. Thirty-six wild cats plus one domestic one are called felid. A member of this family is also called a felid.

1.1. The heart

Heart is one of the vital organs in body as it pumps oxygenated and nutrient-rich blood to different parts of the body. Heart diseases widely affect blood circulation, causing a series of problems. There are two main types of heart diseases: one affecting the heart valves and other heart muscle (**Figure 1**).

Cats with any may be efficaciously managed through nutrition, exercise, nutrition and medication. With the right food and veterinarian advice, a cat can still enjoy a happy and active life.

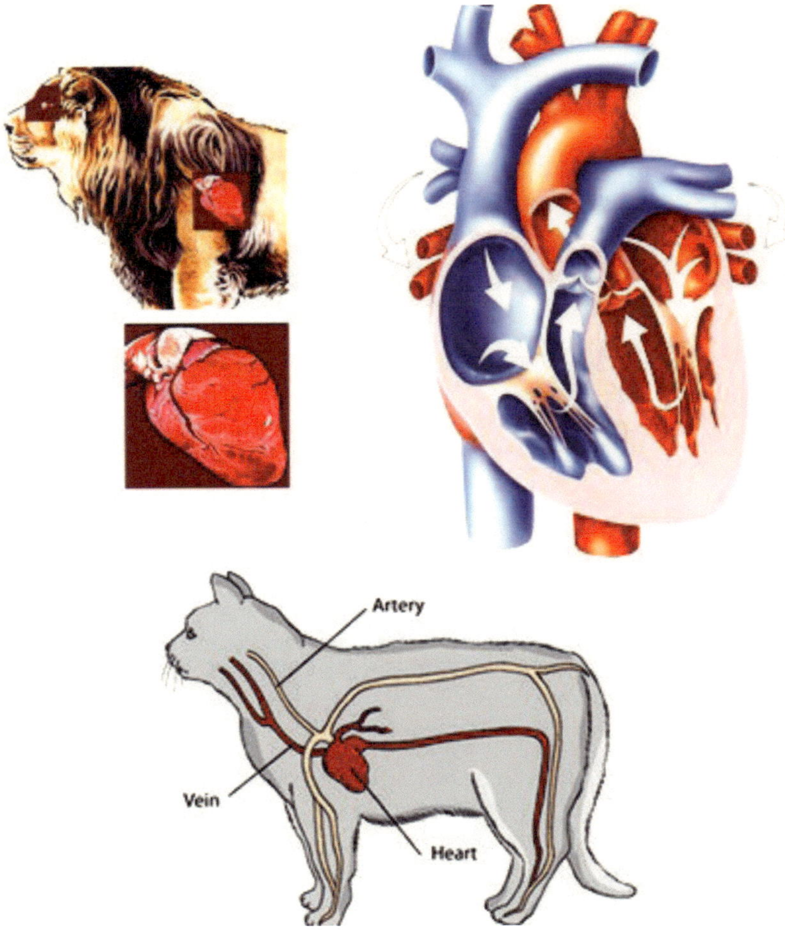

Figure 1. Heart of big and small cats.

1.2. Heart rate

The heart rate ranged between 42 and 76 beats per minute (pulse per minute). Heart charge in large felids is about 40–50 beats consistent with minute. An algometric scaling for critical signs assessment predicts coronary heart price of about 60–80 bpm for mammals weighing 100–250 kg [calculated from the equation: 241 (weight in kilograms−0.25)] [1].

Heart rate of tigers ranged from 56 to 97 bpm. In both types, the most common rhythm was detecting the normal sinus rhythm followed by arrhythmias of the sinus (**Figure 2**). The pacemaker was also observed to wander with normal sinus rhythm or sinus arrhythmias [1, 2].

Veterinary clinical evaluation of the patient with heart disease, especially heart muscle and the deviation from the normal standards, suggests but does not specify structural heart disease. Because deflections and periods of the recordings often are changed by either satisfactory or physiological factors; ECG is a useful tool to diagnose most cardiac arrhythmias and can provide information on the status of the heart muscle. It can also be used as an indicator

http://dx.doi.org/10.5772/intechopen.69781

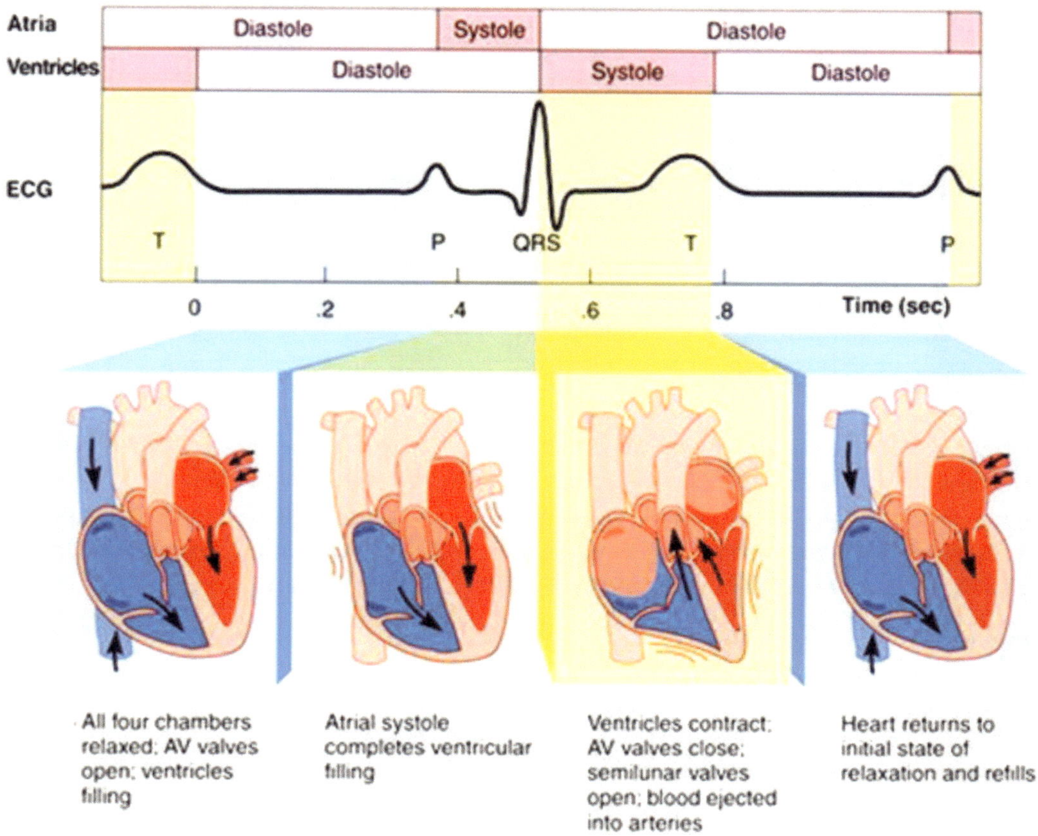

Figure 2. Cardiac cycle. *Heart Rate:* Heart rate of cat ranged from 120 to 140 beats per minute (bpm); heart rate of tigers ranged from 56 to 97 beats per minute (bpm); and heart rate in lion is about 40 to 50 beats per minute. An algometric scaling for vital signs assessment predicts heart rate of approximately 60 to 80 beats per minute (bpm) for mammals weighting 100 to 250 kg (calculated from the equation: 241 [weight in kilograms − 0 25]).

of heart that expands room and electrolyte imbalances. Currently, there is a loss of statistics on the anticipated rhythm and other parameters of electrocardiographic wild felids.

The only source for comparison is data that have been collected from domestic carnivores. Description of every tune recorded at some stage in this study is as common for puppies and cats. It can be attributed to higher values for wave durations found on the largest coronary heart muscles of large felids. A few articles describe invasive oscilloscope and Doppler ultrasound gadgets to measure arterial blood pressure of unusual felids in studies of the usage of diverse anesthetic dealers.

Electrocardiograph (ECG) recordings follow standard procedures as recommended for domestic carnivores. The animals were positioned on the right side lying down, where attached electrodes to the skin via the crocodile moistened clips alcohol in uniform places: the proper arm (RA) and left arm (la) subsequent to the involvement of the caudal aspect of the foreleg appropriate, the proper leg (RL) and left leg (LL) at the patellar ligament on the front facet of the right hind leg (**Figure 3**) [1–3].

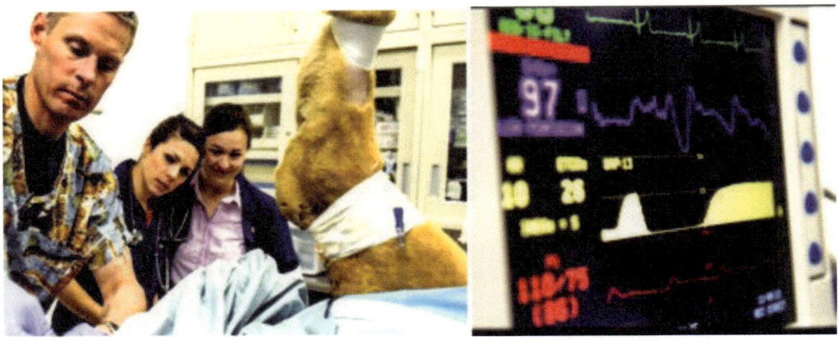

Figure 3. ECG for a lion under anesthesia.

Reports on the use of ECG in wild felines were found intermittently and are usually relevant to assess anesthetic agents. The recorded electrocardiographic parameters of three aware tigers (*Panthera tigris*) have been blanketed rhythms recorded regular rhythm sinus (48.1%), ordinary sinus rhythm with a pacemaker wandering (7.5%), abnormal heart sinus rhythm (18.5%), and cardiac arrhythmia sinus rhythm with a wandering pacemaker (26%).

If the electrical axis between +60 and +90° in 96.3% and in +90° in 3.7% of examinations [1, 2]. In some studies, tigers supplied suggest heart charge of 81 bpm; 38.4% of the animals had ordinary sinus rhythm, 15.4% had normal sinus rhythm with wandering pacemaker, 30.8% had sinus arrhythmia, and 15.4% had sinus arrhythmia with wandering tempo maker. The electrical axis ranged from +60 to +90° in 77.3% of the animals, at +90° in 15.4%, and from +90 to + 120° in 7.3%. Duration of P wave and QRS complex tended to be greater in lions than in tigers and other parameters were similar for each species. This is another wrong fallacy and false notion against the ligers, and ligers do not now have any such problems touching on a weaker coronary heart.

A heart of a liger is as normal as that of the lion and tiger, and there has been never a sign or symptom, which indicates that ligers have a weaker heart. Furthermore, in a study, on ligers where different deaths were analyzed, none of the ligers had died of heart failure. Therefore, it is far entirely incorrect to finish about the ligers that they have got very vulnerable weak spot.

The reason for the relatively low amplitude of the QRS waves, in both species, is to increase the distance between the hearts and re-ropes electrodes. This is due to the fact that the size of the chest and the thickness of the chest wall of the large felids are larger than those of the local carnivores.

2. Hematological parameters

Hematological parameters include hemoglobin (Hgb), hematocrit (Hct), red blood cells (RBC), mean corpuscular volume (MCV), mean corpuscular [8] and hematologic (**Table 1a** and **Figure 4**), and serum chemistry (**Table 1b**) reference intervals for free-ranging lions (*Panthera leo*) [4–8], and other species (**Tables 2–8b**).

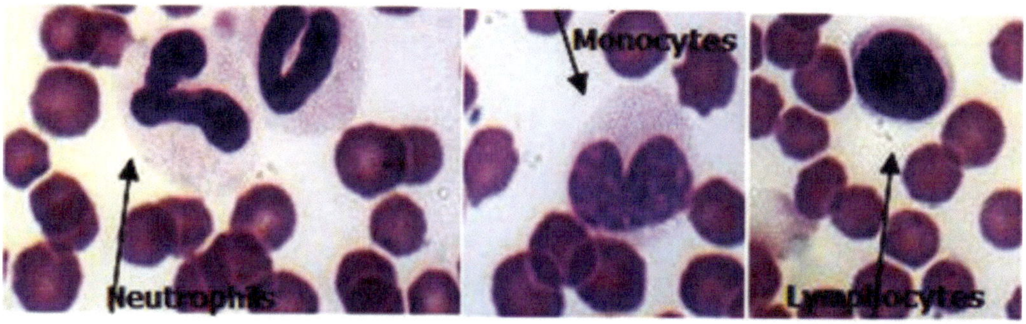

Figure 4. Blood smear of tiger *(Panthera tigris)* stained by Wright stain, 1000×.

Parameter	**Range**
Hgb (g\dl)	8.9–14.6
Hct (%)	26.8–44.1
RBC(×10^6 cells\μl)	5.1–8.3
MCH (pg\cell):	
Male	14.8–18.5
Female	15.5–19.1
MCHC (g\dl):	
Male	29.6–35.5
Female	30.4–35.7
WBC(×10^3 Cells\μl)	7.2–25.6

Table 1a. Means for the hematologic parameters of captive Siberian tigers (*Panthera tigris altaica*) [6–12].

Parameter	**Range**
Cholesterol (mmol\l):	1.8–5.0
Male	26.8–44.1
Female	0.9–6.0
Total proteins (g\l)	66.7–104.2
Albumin (g\l)	20.0–30.0
Globulin (g\l):	
Male	33.9–76.5
Female	40.9–75.1

Table 1b. Means for the serum parameters of captive Siberian tigers (*Panthera tigris altaica*) [6–12].

Parameter	Range
Leukocytes (×10³/ml)	5.0–12.2
Neutrophils (×10³/ml)	70.0–88.0
Lymphocytes (×10³/ml)	10.0–30.0
Monocytes (×10³/ml)	0.0–5.0
Eosinophils (×10³/ml)	0.0–2.0
Bands (×10³/ml)	0.0–2.0
Erythrocytes (×10⁶/ml)	4.3–7.8
Hemoglobin (g/dl)	8.4–14.0
Packed cell volume (%)	26.5–46.0
Mean corpuscular volume (fl)	56.0–62.0
Mean corpuscular hemoglobir concentration (g/dl)	29.0–31.8
Platelet estimate (×10³/ml)	336.0–502.0
Albumin (g/dl)	3.3–3.9
Alkaline phosphatase (IU/l)	14.0–57.0
Bilirubin (mg/dl)	0.1–0.2
Blood urea nitrogen (mg/dl)	28.0–46.0
Calcium (mg/dl)	8.8–10.1
Cholesterol (mg/dl)	92.0–122.0
Chloride (mEq/l)	19.0–126.0
Creatinine (mg/dl)	1.5–3.3
Gamma-glutamyl trans peptidase (IU/l)	0.0–1.0
Glucose (mg/dl)	79.0–136.0
Potassium (mEq/l)	3.8–4.9
Sodium (mEq/l)	55.0–158.0
Phosphorus (mg/dl)	3.9–6.5
Protein (g/dl)	6.4–7.6
Aspartate amino-transferase (IU/l)	20.0–32.0
Alanine amino-transferase (IU/l)	23.0–46.0
Thyroxine (mcg/dl)	1.8–2.9

Table 2. Hematologic and serum biochemistry values of captive lynx.

Blood constituents (cells, plasma, and chemical composition) can be used to display the fitness reputation and diagnose diseases, dietary deficiencies and the reproductive popularity (i.e., pregnancy) of animals. Techniques to perform complete blood evaluation are to be had for humans and maximum farm animals' species. The same processes can be used for wildlife species to perceive deviations from normal values in positive blood parameters (cytological or biochemical).

Parameter	Mean
PCV (%)	41.5 ± 5.1
Hb (g/dl)	12.1 ± 1.1
RBC (×10^{12}/l)	6.9 ± 1.0
MCV (fl)	63.1 ± 7.0
MCHC (g/dl)	30.5 ± 3.6
WBC (×10^9/l)	7.07 ± 1.76
Bands (×10^9/l)	0.18 ± 0.12
Segmented neutrophils (×10^9/l)	4.44 ± 1.44
Eosinophils (×10^9/l)	0.41 ± 0.23
Basophils (×10^9/l)	0.01 ± 0.01
Lymphocytes (×10^9/l)	1.53 ± 0.48
Monocytes (×10^9/l)	0.32 ± 0.18
Bands (%)	2.0 ± 1.5
Segmented neutrophils (%)	64.8 ± 8.8
Eosinophils (%)	6.0 ± 4.0
Basophils (%)	0.2 ± 0.2
Lymphocytes (%)	22.2 ± 5.3
Monocytes (%)	4.6 ± 3.4
Reticulocytes (%)	0.67 ± 0.48
Plasma protein (g/dl)	8.3 ± 1.2

Table 3. Hematologic values of fishing cat (*Felis viverrina*).

Hematology	Unit	Range
Red blood corpuscles	(TEC) × 106/μl	4.66–9.15
Total leukocytes count	(TLC) × 103/μl	6.2–11.05
Hemoglobin	(Hb) g/dl	7.8–13.8
Hematocrit	(PCV) Ratio	36–45
Erythrocyte sedimentation rate	(ESR) Hours	14–26
Icterus index	(II) u/l	2–5
Differential leukocyte count	%	
Neutrophils		57–75
Lymphocytes		18–35
Monocytes		2–6
Eosinophils		2–6

Hematology	Unit	Range
Basophils		0–4
Blood plasma biochemistry		
Albumin (ALB)	g/dl	2.1–4.6
Total protein (TPROT)	g/dl	3.7–8.7
Total bilirubin TBIL)	mg/dl	0.4–3.2
Creatinine (CRE)	mg/dl	1.6–4.6
Blood urea nitrogen (BUN)	mg/dl	6.5–48.2
Alanine aminotransferase (ALT)	IU/l	21.2–109.0
Aspartate aminotransferase (AST)	IU/l	14.4–84.0

Table 4. Hematological and biochemical values of Bengal tigers *(Panthera tigris tigris).*

Parameter	Unit	Range
PCV	%	30–45
Hgb	g/dl	9.8–15.4
RBCs	$\times 10^6/\mu l$	5.0–10.0
Reticulocytes	%	0–0.6
Absolute reticulocyte count	$\times 10^3/\mu l$	<60
MCV	fl	39–55
MCH	pg	13–17
MCHC	g/dl	30–36
Platelets	$\times 10^3/\mu l$	300–800
MPV	fl	12–18
WBCs	$\times 10^3/\mu l$	5.5–19.5
Neutrophils	%	45–64
Cholesterol	mg/dl	71–156
Creatinine	mg/dl	0.9–2.2
Glucose	mg/dl	60–120
Magnesium	mg/dl	1.7–2.6
Phosphorus	mg/dl	3.0–6.1
Potassium	mEq/l	3.7–6.1
Protein	g/dl	6.0–7.9
Albumin	g/dl	2.8–3.9
Globulin	g/dl	2.6–5.1

Parameter	Unit	Range
Sodium	mEq/l	146–156
Urea nitrogen	mg/dl	19–34
ALT	U/l	25–97
Amylase	U/l	550–1458
Alk phos	U/l	0–45
AST	U/l	7–38
CK	U/l	69–214
GGT	U/l	
LDH	U/l	58–120
SDH	U/l	
Bicarbonate	mEq/l	17–24
Bilirubin	mg/dl	0–0.1
Calcium	mg/dl	8.7–11.7
Chloride	mEq/l	115–130

Table 5. Hematologic and serum chemistry values of domestic cat (*Felis catus*).

Hematological parameter	Mean	
Erythrocytes ($\times 10^6/\mu l$)	8.97	
Hematocrit (%)	42.38	
Hemoglobin (g/dl)	14.11	
MCV(fl)	47.70	
MCH (pg)	15.84	
MCHC (%)	33.33	
Leukocytes ($\times 10^3/\mu l$)	9.37	
Differential cell count (cells/μl)		
Band neutrophils	0	
Mature neutrophils	7.748	
Eosinophils	372	
Basophils	4	
Lymphocytes	849	
Monocytes	365	
Plasma fibrinogen (g/dl)	0.23	
Serological parameters	***Range***	
Sodium (mmol/l)	133.0	161.0

Hematological parameter	Mean	
Chlorine (mmol/l)	108.8	133.3
Total Bilirubin (μmol/l)	2.0	8.0
Total carbon dioxide (mmol/l)	7.7	18.6
Gamma-glutamyl transferase (IU/l)	0.0	4.0
Uric acid (mmol/l)	0.0	0.1
Albumin (g/l)	9.0	14.0
Total Direct Bilirubin (μmol/l)	0.0	2.0
Aspartate transaminase (IU/l)	12.0	40.0

Table 6. Hematologic and serum biochemistry parameters of captive lions (*Panthera leo*).

Parameter	Units	Mean
Red blood cells (RBC)	$\times 10^6/\mu l$	7.635
Hemoglobin (Hb)	g/dl	12.21
Packed cell volume (PCV)	%	36.37
Mean cell volume (MCV)	fl	47.29
Mean cell hemoglobin (MCH)	pg	16.07
Mean cell hemoglobin conc. (MCHC)	g/dl	34.08
Red cell distribution width (RDW)	fl	22.12
Reticulocytes	% RBC's	1.04
Nucleated RBC's (NUC)	/100 RBC's	1.5
White blood cells (WBC)	$\times 10^3/\mu l$	12.19
Segmented neutrophils	$\times 10^3/\mu l$	8.0
Segmented neutrophils	% WBC's	64.3
Lymphocytes	$\times 10^3/\mu l$	3.4
Lymphocytes	% WBC's	28.8
Monocytes	$\times 10^3/\mu l$	0.39
Monocytes	% WBC's	3.2
Basophils	$\times 10^3/\mu l$	0.10
Basophils	% WBC's	0.89
Eosinophils	$\times 10^3/\mu l$	0.42
Eosinophils	% WBC's	3.4
Platelets	$\times 10^3/\mu l$	402.6

Table 7a. Hematologic parameters of jaguar.

Parameter	Units	Mean
Albumin	g/dl	3.70
Alanine aminotransferase (ALT/SGPT)	U/I	60.2
Alkaline phosphatase (ALP)	U/I	35.4
Aspartate aminotransferase (AST/SGOT)	U/I	73.4
Calcium (Ca)	mg/dl	9.92
Carbon dioxide (CO_2)	mEq/l	14.33
Cholesterol	mg/dl	147.9
Chloride (Cl)	mEq/1	115.5
Creatine phosphokinase (CPK)	U/I	515.6
Creatinine (Creat)	mg/dl	1.84
Gamma glutamine transferase (GGT)	U/I	1.6
Glucose (Gluc)	mg/dl	154.4
Inorganic phosphorus (IPhos)	mg/dl	5.77
Iron (Fe)	µg/dl	65.1
Lactate dehydrogenase (LDH)	U/I	269.7
Potassium (K)	mEq/l	4.60
Sodium (Na)	mEq/l	152.6
Total bilirubin (Tbili)	mg/dl	0.26
Total protein (TP)	g/dl	7.35
Triglycerides (Trig)	mg/dl	54.9
Urea nitrogen (UN)	mg/dl	37.7
Uric acid	mg/dl	0.55

Table 7b. Serum and biochemistry parameters of jaguar.

Parameter	Unit	Range
Hematology		
Hemoglobin	g/dl	3.6–14.8
Hematocrit	%	12.9–47.7
Red blood cells	10^6/µl	2.72–10.6
Mean cell volume	fl	43.0–54.0
Mean cell hemoglobin	pg	13.2–17.0
Mean cell hemoglobin conc.	g/dl	27.9–34.4

Table 8a. Hematologic parameters in *Puma concolor couguar*.

Parameter	Unit	Range
Serum chemistry		
Glucose	mg/dl	36.0–258.0
Urea nitrogen	mg/dl	17.0–78.0
Creatinine	mg/dl	0.5–3.4
Total protein	g/dl	5.7–8.7
Albumin	g/dl	1.1–4.1
Total bilirubin	mg/dl	0.1–0.3
Alkaline phosphatase	U/I	1.5–24.0
Alanine aminotransferase	U/I	19.0–215.0
Aspartate aminotransferase	U/I	27.0–465.0
Cholesterol	mg/dl	51.0–217.0
Calcium	mg/dl	7.8–10.9
Phosphorus	mg/dl	2.1–10.9
Sodium	mEq/l	145.0–168.0
Potassium	mEq/l	3.6–7.2
Chloride	mEq/l	96.0–124.0
Globulin	g/dl	2.7–6.1
Lipase	U/l	0.0–93.0
Amylase	U/l	87.0–612.0
Triglycerides	mg/dl	4.0–283.0
Creatine phosphokinase	U/I	133.0–1332.0
γ	Glutamyl transpeptidase GGT	U/I
Magnesium	mEq/l	1.4–9.8
Osmolality	mOsm/l	292.0–365.0

Table 8b. Serum and biochemistry in *Puma concolor couguar*.

Values of blood parameters in large cats (**Tables 1a** and **1b**) can be used by veterinarians or researchers to assess animal health or population. The blood parameter values may vary between free ranging and captive species [4, 5–12].

3. Cardiac diseases in Felidae

3.1. Arrhythmogenic right ventricular cardiomyopathy

The cause of this disease is unknown and results from a gradual atrophy of the right ventricular myocardium replaced with fibrous and/or fatty, mirroring the disease counterpart in humans [12, 13].

Arrhythmogenic right ventricular cardiomyopathy has mentioned family inclinations for ARVC in massive cats, but it wishes in addition research.

It detected apoptosis in a high percentage of cat's hearts ARVC, with mean thoughts similar to those reported in patients with human ARVC index. This conclusion is supported by the hypothesis that causes the pathogenesis of ARVC, may depend in part on inflammation as well as apoptosis. Coexistence between the cells and the inflammation of the heart muscle in cats with ARVC indicates that these mechanisms contribute to muscle injury and repair in felines likely indicate that the heart muscle inflammation may represent a starting point for apoptosis. In most cats, death is a common sequel by the time clinical signs become apparent due to congestive heart failure right side gradually. And it is often misdiagnosed as cats ARVC tricuspid dysplasia of honor, but the latter always happens disease in cats as young congenital abnormality, which is not true deep cardiac hypertrophy treatment. There is no justification for the mechanical removal of pleural effusion by thoracentesis or closed thoracostomy tube when hoarseness exists. Stabilization of cat cases includes a scientific technique to use an angiotensin-converting enzyme inhibitors, diuretics, and digoxin, with variable achievement.

Antioxidants such as furosemide (2–4 mg/kg/day), spironolactone (5–10 mg/Eg fortekor 2.5 mg/cat daily), and digoxin (1/4 0.125 mg oral tablet P 24 hours).

3.2. Atrial fibrillation

Cat's heart consists of four chambers [13]. Higher chambers are called atria (one: the atrium), and the lower chambers are called ventricles.

Valves are present between each pair of atria and ventricles, each on the left and right hand sides. The coronary valve is a triplex valve found between the left atrium and the left ventricle. Constant rhythmic pattern is a result of an exceptional synchronization and harmony found between numerous atrial and ventral structures. Loss of this synchronization results in several disorders like atrial flutter and atrial fibrillation. Atrial fibrillation is a condition that delays electrical conduction from the atria to the ventricles, or for a prolonged period on ECG, this suggests that the PR period is prolonged time between the principle electric impulses, referred to as the P wave, the QRS complex, which is identified because the coronary heart beats.

Atrial fibrillation (arrhythmia) often arises from atrial flutter. Atrial flutter is characterized by a premature electrical impulse originates in the atria, causing a rapid heartbeat with either regular or irregular frequency.

Arrhythmia or atrial fibrillation results in irregular rhythms of the ventricle as well. On the electrocardiogram, which measures the electrical activity of the heart, a clear pattern can be discerned in atrial fibrillation and atrial flutter.

Atrial fibrillation is classified by:

- Primary atrial fibrillation
- No underlying heart disease involved—caused has not been identified
- Secondary atrial fibrillation
- Severe underlying heart disease such as CHF is involved
- Atrial fibrillation
- Frequent cyclic attacks persist for a short period of time (less than seven days) as the heart returns to normal rhythm on its own
- Continuous atrial fibrillation
- Irregular heartbeat continues (arhythmia) for more than 48 hours, and only responds to treatment
- Permanent atrial fibrillation
- Irregular, nontreatable arrhythmia symptoms are generally associated with underlying disease such as congestive heart failure (CHF).

The symptoms related to atrial fibrillation are (**Figure 5**):

- Heart stimulation(galloping heart)
- Vulnerability
- Coughing
- Shortness of breath (difficulty breathing)
- Tachypnea (rapid breathing rate)

Figure 5. Symptoms of cardiac affections. Weakness, fainting, and syncope/loss of consciousness.

- Idle (lethargy)
- Chronic heart disease which involves valves
- Syncope/loss of consciousness (rare)
- Myocardial infarction (heart muscle disease)
- Expanding the heart(enlargement)
- Congenital heart disease
- Tumors digoxin (drugs commonly used to treat various heart disease (toxicity). As a sequel congestive heart failure, cause may remain unknown.

3.2.1. Diagnosis

- Taking a detailed history of the health of the cat, and the onset of symptoms and potential accidents.
- Veterinarian conducts a full physical examination.
- Lab tests will include full blood tests, a glimpse of biochemistry, and urinalysis.
- It is possible that the results of these tests may not reveal a lot of information related to this disease, but it may be useful to get a comprehensive picture of the health of the cat and the detection of other diseases.
- Additional diagnostic tools echocardiography, X-ray imaging, and color Doppler. The treatment will be directed toward normalizing heart rhythm and get a sinusoidal atrio-ventricular node back into sync with the atrioventricular node (AV) node. If fibrillation is a chronic problem, the success rate drops accordingly. ECT can be used to normalize the rhythm in some cases. If coronary heart disease at the back of, including the Swiss franc gift, it will also be directed closer to the treatment of treatment, at the side of the success of stability in a heartbeat.

3.3. Atrioventricular heart block

Usually, the reason is the contraction of the heart through electrical impulses that arise from the node, the sino-atrial, and stimulate the atria, and travel to the atrioventricular node, and finally to the ventricles [12, 13]. This electrical conduction system is responsible for controlling heart rate, generating electrical pulses (waves), which was published through the heart muscles, and stimulates the heart muscles to contract and push blood through the interior of the arteries in and out of the body (**Figure 6**).

3.3.1. First-degree atrioventricular block

First degree AV block is common in young felines and healthy cats as a result of high vagal tone (impulses generated by the vagus nerve causing inhibition of heart beat) or accompanied

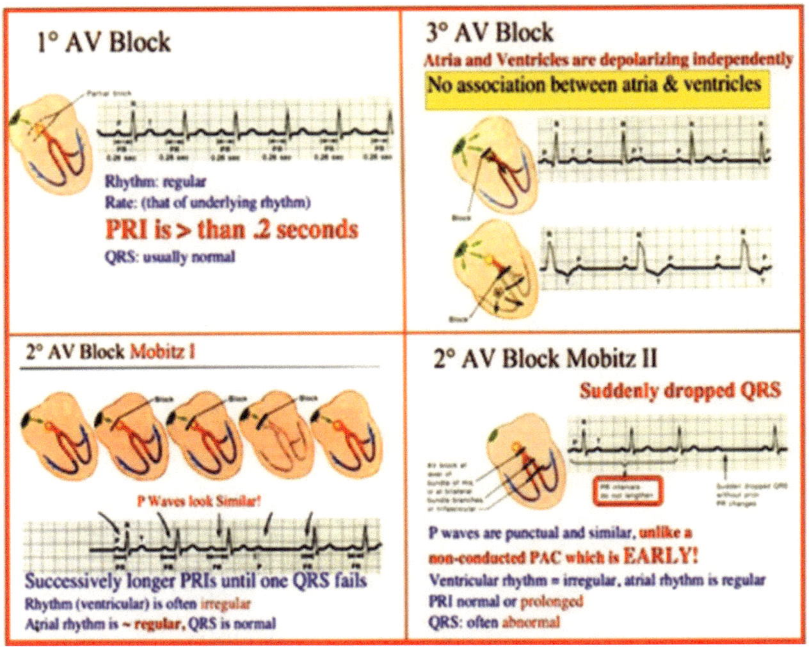

Figure 6. Degrees of atrioventricular heart block in Felidae.

with degenerative conduction system disorder, which by order activates the AV node, conducting regular electrical impulses from the atria to the ventricles with mechanical activity to force blood constantly into the ventricles just before the ventricles are activated to contract and push blood to the body using the pulmonary artery and the aorta?

Affected cases may present variable symptoms

- Congestive heart failure manifestations conjoined with hypertrophic cardiomyopathy
- Shortcoming and overall weakness
- Healthy cats are also prone
- Disorders indirectly related to the heart
- Excessive production of thyroid hormone (hyperthyroidism)
- Cardiomyopathy
- Heart neoplasia
- Drug interference with the function of the AV node

3.3.1.1. Treatment

Treatment will vary depending on the underlying disease causing the second-degree Mobitz Type 1 atrioventricular block. A pacemaker may be necessary.

3.3.2. *Second-degree AV block*

A disease in which the previously stated electrical conduction system goes off track, where part of impulses is conducted down from the atria to the ventricles, causing failure of the heart muscle to contract and pump blood efficiently.

3.3.2.1. *Symptoms and types*

- Weakness
- Idle
- Sudden collapse
- Fainting (fainting) causes involving noncardiac diseases
- Age-related degenerative changes within the cardiac delivery system side effects of drugs (e.g., digoxin, a drug used to treat many heart diseases)
- Heart tumors
- Heart-related infections (e.g., bacterial, viral, parasitic)
- Myocardial infarction (myocardial infarction)
- Shock or trauma

3.3.3. *Third degrees of atrioventricular heart block*

3.3.3.1. *Diagnosis*

The precise date for the health of the cat is important, and the onset of symptoms, and possible incidents that preceded this condition. After a full physical examination, the cat arterial blood pressure measurement makes sure that high blood pressure (hypertension) is associated with heart disease. The laboratory test includes standard full blood count, personal biochemistry and urine analysis. These tests are important in diagnosing this problem as there are some biochemical changes that can create a cat to prevent AV block, for example, the presence of an infectious disease or parasitism tests. The culture/sensitivity of the blood test guide will be shown for the type of organisms involved in the infection and the sensitivity of various antibiotics. Other diagnostic tools which are important for comparing structural and useful cardiac parameters are encompass ECG (EKG) and echocardiography for measuring electrical pulses.

3.3.3.2. *Treatment*

Do not treat the disease aggressively in cats. If it is to maintain the heart rate at the level at which the heart can pump enough blood to the body's natural functions, and generally will be required to treatment. If the primary disease is responsible for AV block, and the veterinarian to deal with them accordingly.

3.4. Cardiac tumors

Body aorta and carotid tumors, recognized as chemodectomas, mostly are benign tumors (**Figure 7**) that arise from chemical receptor tissues in the body. Oxygen content and PH levels in the blood are highly labile to body's chemical changes. Chemical receptors are present over a wide range in the body; however, chemical receptors in the aorta and carotid artery systems are the main ones affected by chemodectomas.

Chemodectomas are rare in cats, but when they occur, older cats tend to be more predisposed. However, it does not seem to be sex or breed a mile of chemodectomas. Given that this is a rare condition in cats, and aortic tumors are more common than carotid tumors, but a malignant tumor in the other organs seem to be more common in cats when it does not happen.

Symptoms and types associated with aortic body tumors include:

1. Coughing
2. Trouble breathing
3. Symptoms of right-sided congestive heart failure
4. Weakness, lethargy carotid body tumors, near the bifurcation point on the carotid artery, where arteries of internal and external carotid originate. The main function of these arteries is to transport oxygenated blood to upper organs like head and neck. Owing to complicity of this area and the function of theses arteries, carotid tumors are often inoperable.

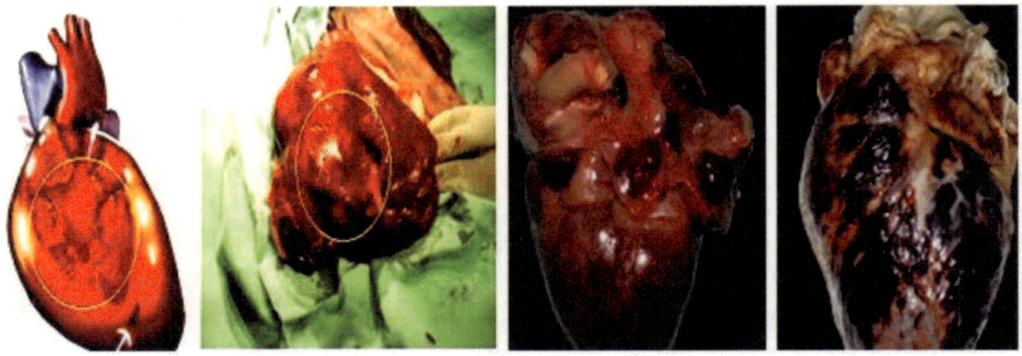

Figure 7. Aortic and carotid body tumors, classified as chemodectomas, are generally benign tumors that grow from the chemoreceptor tissue of the body.

In maximum cases, these tumors nonetheless gradually grow, but benign, as is the case with aortic tumors, they end up a health hassle after they invade the spaces of neighboring blood vessels and lymph vessels. In an expected 30% of the instances, a malignant tumor within the surrounding organs, along with lungs, bronchia or lymph nodes, or more in the liver or pancreas, may occur:

http://dx.doi.org/10.5772/intechopen.69781

1. Gastroesophageal reflux
2. Vomiting
3. Eating disorders (anorexia)
4. Neck lumps
5. Severe hemorrhaging due to presence of tumors in blood vessels (sudden death may occur)
6. Up to 50% of the cases show metastasis to local blood vessels.
7. Up to 20% of the cases show organ failure due to cancerous growths.

3.5. Chronic valve disease hypertrophic cardiomyopathy (HCM)

It is a disease that is maximized (enlarged) part of the heart muscle (myocardium) without any apparent reason and the creation of a functional disorder of the heart [12, 13]. This is the main cause of sudden death. The prevalence of hypertensive cardiomyopathy is the main cause of sudden cardiac demise in any age organization and the reason of heart attack signs. HCM is often asymptomatic until sudden death. This is frequently one point missense mutations inside the genes of beta-myosin heavy chain (MHC), the myosin-binding protein C, troponinT coronary heart, or tropomyosin. These mutations reason muscle and muscle structural malformations and ability defects in power generation.

Myocardial infarction is a disease that affects my heart muscle. With them, myosites (heart-limiting cells) increase the heart's volume, leading to cardiomyopathy. In addition, the natural alignment of muscle cells is disrupted, a phenomenon known as myocardial disarray and dysfunction (**Figure 8**).

Manifestations of the clinical course of HCM are fluctuating. A number of cases are asymptomatic or showing mild symptoms. Symptoms incorporate dyspnea (short breath) owing to stiffening and reduced blood filling of the heart ventricles, chest pain with effort and exertion (angina) as a result of decreased or restrained flow of the blood (referred to as ischemia) to the coronary arteries, palpitations due to the preceded ischemia, in addition to the interruption of the electrical system running across the defected heart muscle, light headedness, lethargy, passing out (Syncope) and sudden cardiac arrest. Respiratory distress basically attributed to accelerated left ventricular rigidity, which weakens the process of filling the ventricles, and however, also results in better strain within the left ventricle and atrium, resulting in posterior stress and interstitial congestion of the pleura.

Manifestations are distantly related to the existence and severity of the outermost gradient. Symptoms often imitate those of congestive heart failure (especially in activity and short of breath), but with different therapeutic regimen. Beta-blockers are used in each case but, using diuretics as a major pillar in therapy, will aggravate the signs and manifestations of obstructive hypertensive myocardial infarction through reducing the scale of the ventricular load and accordingly growing in the glide resistance (less blood to push aside thick blocking tissue).

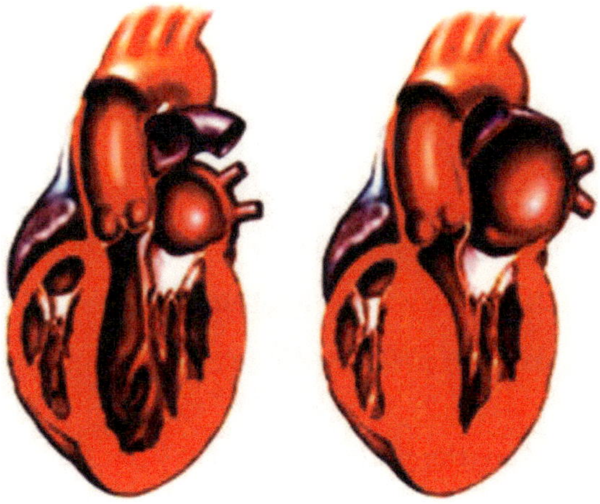

Figure 8. HCM also causes disruptions of the electrical functions of the heart. The cardiomyopathy is a disease that affects the heart muscle, with HCM and myocytes (the contractile cells of the heart) to increase the heart rate in size, leading to a thickening of the heart muscle. In addition, a phenomenon, the natural alignment of the cells of muscle breakdown, is known as chaos of the heart muscle. HCM also causes disruption of electrical functions of the heart.

The main risk of sudden death factors in individuals with an earlier history of cardiac arrest or ventricular fibrillation, spontaneous arrhythmia constant heart beat and ventricle, in exercise, blood pressure tachycardia and irregular heart rate and ventricle.

3.6. Dilated cardiomyopathy

Enlarged coronary heart (dilated cardiomyopathy) in cats is a heart disorder that impacts the ventricular muscle [12, 13].

It is characterized by dilated heart chambers, or enlarged, reduced the ability of deflation, that is, the lower the ability to push blood from the ventricle of them. It is suspected that this may have been related to the food shortages of the amino acid taurine.

Manifestations of reduced cardiac blood flow, caused by DCM:

- Physical and mental suffering (distress and depression), anorexia, and lethargy.
- Reduced blood flow as a result of blocked blood vessels, thromboembolism can be readily seen as an abrupt onset of ache and paraparesis (referred to as partial paralysis).
- Further examinations can show low or high coronary heart charge or regular, soft coronary heart puff, drop rhythm, hypothermia, left heart weak point, and quiet lung sounds.

3.6.1. Diagnosis

In addition to a comprehensive physical examination of the heart, some medical tests are needed for the diagnosis of DCM and the exclusion of other diseases. To record the electrical (or EKG) can be used to study electrical currents in the heart muscle, can reveal any defect in the electrical conduction of the heart (which lies behind the ability of the heart to contract/pulse), and can also help the veterinarian to determine the origin of irregular heartbeat, if they are present. X-ray imaging of the chest (pectoral rays) may reveal enlarged heart and fluid accumulated in the chest. Required echocardiography (ultrasound) imaging is carried to confirm the diagnosis of DCM. This test enables the veterinarian to check the eyesight of heart size and capacity of the ventricular muscle to contract. And echocardiography may reveal the thin walls of the ventricle, which is an enlarged left ventricle and the left atrium, and a reduced ability of deflation, which confirms the presence of the diagnosis of DCM. Treatment DCM varies with the state of the cat. If cat is suffering from severe symptoms, hospitalization may be necessary. It may include treatment for DCM drugs to control irregular heartbeat, and the Department of Health to prevent renal failure, treatment of low blood pressure, and treatment of complications from blood clots (i.e., blood thinning drug). And treatment in the hospital for congestive heart failure usually includes supplemental oxygen therapy, diuretics to reduce fluid retention, nitroglycerin to improve blood flow, and low doses of dobutamine to stimulate cardiac contractility and cardiac output. Other drugs, such as anticoagulants (blood) liquidity, and beta-blockers to control the rhythm can be used for the treatment of DCM, but their use depends on the specific problems that are secondary to the disease. Cats suffering from DCM will typically have loss of appetite, but they also need to give a low-sodium diet, to reduce the tension of fluid on the heart, and you need a diet that would raise the interesting cat eating plan, in order to assist in its recovery.

3.7. Pulmonary hypertension

Pulmonary hypertension happens in feline due to constriction of the pulmonary arteries and capillaries, blockage, or gain superfluous blood flow. Blood capillaries in the lung are minute branches of blood vessels, approximately thick as one cell. Linking smaller veins into smaller arteries for oxygen and carbon dioxide swapping between blood and surrounding tissues. Arteries transport oxygenated blood from the heart to the lungs, and pressure exerted by the blood on the left atrium can result in high pressure in the capillaries in the lungs.

3.7.1. Symptoms and types

1. Exercise intolerance
2. Trouble breathing
3. Bluish-purplish tinged skin
4. Coughing
5. Coughing or vomiting up blood
6. Enlarged abdomen
7. Weight Loss
8. Fatigue
9. Fainting

3.8. Pulmonic stenosis

Pulmonic stenosis is rarely seen in cats. Pulmonary stenosis is a congenital heart defect (the present one since birth), which causes a narrowing of the pulmonary valve to the heart (**Figure 9**). Pulmonary valve connects the right ventricle heart chamber to the pulmonary artery that carries blood from the heart to the lungs. When the pulmonary valve stenosis, in blocking the flow of blood from the heart to the lungs, right ventricle must work harder than usual to push the blood through. In many cases, the pressure builds up behind the obstructed pulmonary valve, leading to accumulation of fluid in the abdomen or chest that leads to congestive heart failure. Overworking the heart can also cause a heart murmur (abnormal heart sound) and cardiac arrhythmia (abnormal heart rhythm).

The defect or the illness is generally observed, while a heart murmur is detected all through a routine checkup, warranting further investigation. In moderate or severe cases, the animal may show symptoms of congestive heart failure. It includes:

- Difficulty breathing/shortness of breath
- Weakness
- Exercise intolerance
- Collapse treatment for pulmonic stenosis will ultimately depend on the severity of the defect. Cats with light or mild cases may live normally without the need for intervention. Cats with light or moderate cases may additionally live normally without the need for intervention. Cats with slight-to-severe defects, however, are normally handled with a minimal surgery called a vulvoplasty valve balloon. Balloon valvuloplasty is a catheterization

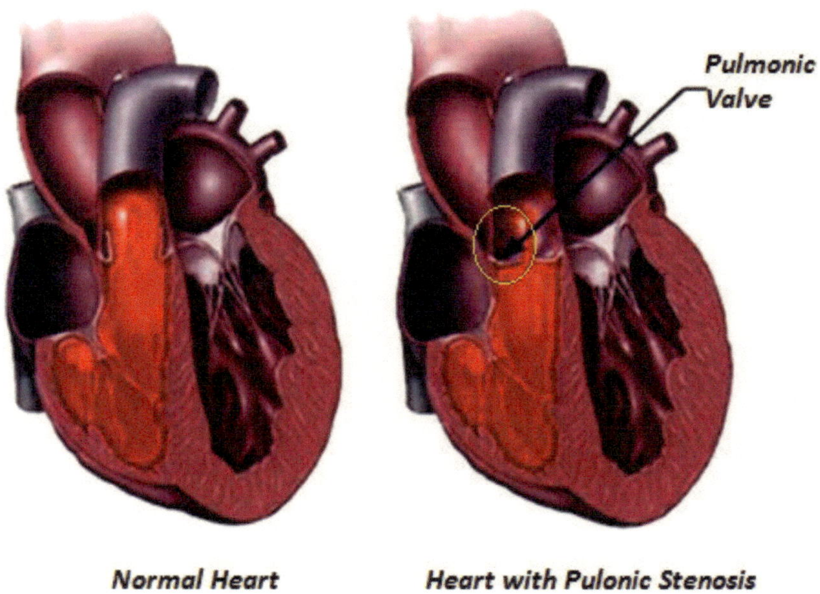

Figure 9. Pulmonic stenosis can range from mild to severe.

procedure, wherein a balloon is guided to the narrowed a part of the valve after which inflated, causing the valve to stretch and blood flow to improve.

3.9. Subaortic stenosis

- A rare defect in the cat [12, 13].
- Made of abnormal tissue placed under the aortic valve causing a blockage to the coronary artery. The cardiac muscle must exert more effort to pump the blood toward the frame.
- As a result of exerting more effort, the cardiac muscle starts to thicken (hypertrophy). Blood pumped at an elevated speed and pressure than normal through the stenosis into the aorta and creates heart murmur.

3.9.1. Symptoms

1. Lethargy.
2. Weakness following exercise or excitement.
3. Fainting.
4. In some advanced cases, coughing and difficulty breathing secondary to congestive heart failure.

Diagnosis of SAS echocardiogram with Doppler under supervision of board-certified veterinary specialized in heart disorders is a powerful tool helping to visualize the atria and ventricles, in addition to the subaortic area. Using Doppler helps in estimating the strain present in the heart by blockage (stenosis).

There is a correlation between elevated pressure and the degree of SAS. Cases with irregular cardiac rhythm need further examinations using ECG (**Figure 10**).

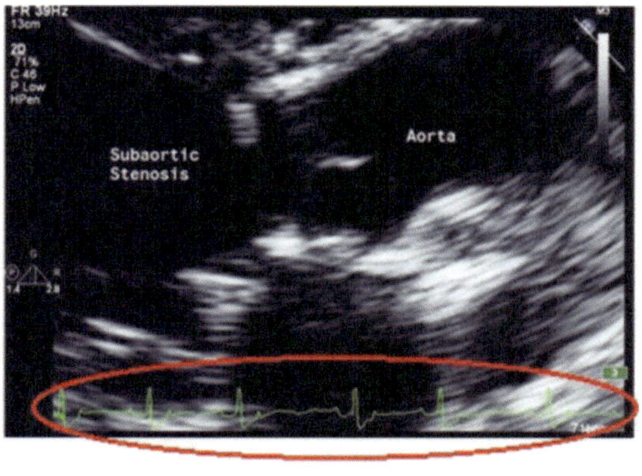

Figure 10. Echocardiogram with Doppler.

Treatment:

1. Prophylactic antibiotics

2. Limited exercise

3. Cardiac medications:

- Beta-blockers are often recommended for moderate-to-severely-affected SAS
- Additional therapy to treat specific arrhythmias and heart failure may also be required.

3.10. Left ventricular noncompaction

Left ventricular noncompaction or "spongy myocardium" is a rare congenital cardiomyopathy that can be diagnosed at any age. Eventually, this condition can potentially lead to chronic heart failure, life-threatening ventricular arrhythmias, and systemic embolic events. LVNC is a condition of the heart where the walls of the left ventricle (the bottom chamber of the left side of the heart) are noncompacted. This causes channels to form in the heart muscle, called trabeculations (**Figure 11**).

This offers the left ventricle a feature 'sponge' feature (a touch like honey). Even though it commonly impacts the left ventricle, it can additionally affect the proper ventricle.

Symptoms of left ventricular noncompaction (LVNC):

- Breathlessness
- Fatigue (extreme tiredness)
- Feeling dizzy
- Fainting or passing out (Syncope).

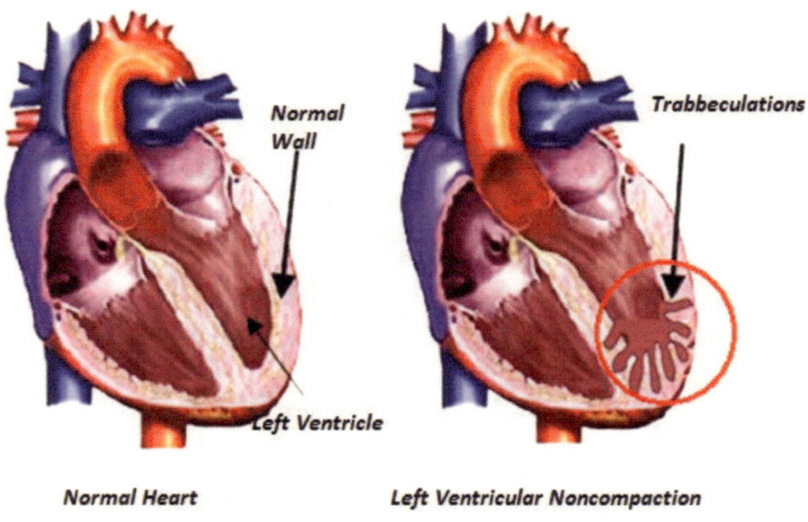

Figure 11. Left ventricular noncompaction or "spongy myocardium," is a rare congenital cardiomyopathy that can be diagnosed at any age.

http://dx.doi.org/10.5772/intechopen.69781

Author details

Hussein Abdelhay Kaoud

Address all correspondence to: ka-oud@outlook.com

Department of Hygiene and Management, Faculty of Veterinary Medicine, Cairo University, El-Giza, Egypt

References

[1] Larsson MHMA. Electrocardiographic parameters of captive lions (*Panthera leo*) and tigers (*Panthera tigris*) immobilized with ketamine plus xylazine. Journal of Zoo and Wildlife Medicine. 2008;**39**(3):314-319

[2] Yamaki FL. Padronização do monitoramen to eletrocardiografico ambulatorial em 24 horas (sistema Holter) em gatos domesticos clinicamente normais [Dissertação (Mestrado)]. Sao Paulo, Brazil: Faculdade de Medicina Veterinaria e Zootecnia, Universidade de Sao Paulo; 2001

[3] [Internet]. 2017. Available from: http://www.comstocksmag.com/snap/tall-orderIn:2017 San Diego Zoo Global Wildlife Conservancy (Zoo Animal and Wildlife Immobilization and Anesthesia)

[4] Hawkey CM, Hart MG. Haematological reference values for adult pumas, lions, tigers, leopards, jaguars and cheetahs. Research in Veterinary Science. 1986;**41**:268-269

[5] Teare JA, editor. International Species Information System (ISIS) Physiological Data Reference Values. CD-ROM. Apple Valley, USA; 2002. [Electronic resource], HYPERLINK "http://www.nla.gov.au/libraries" Australian Libraries Gateway

[6] Maas M, Keet DF, Rutten VPMG, Heesterbeek JAP, Nielen M. Assessing the impact of feline immunodeficiency virus and bovine tuberculosis co-infection in African lions. Proceedings of the Royal Society B: Biological Sciences. 2012;**279**:4206-4214

[7] Wack RF. Felidae. In: Fowler ME, Miller RE, editors. Zoo and Wild Animal Medicine. Missouri: Saunders; 2003. pp. 491-501

[8] Larsson MHMA. Hematologic parameters of captive lions (*Panthera leo*) and Siberian tigers (*Panthera tigris altaica*). Acta Scientiae Veterinariae. 2015;**43**:1311

[9] Currier MJP, Russel KR. Hematology and blood chemistry of the mountain lion *Felis concolor*. Journal of Wildlife Disease. 1982;**18**(1):99-104

[10] Dunbar MR, Nol P, Linda SB. Hematologic and serum biochemical reference intervals for Florida panthers. Journal of Wildlife Disease. 1997;**33**(Suppl 4):783-789

[11] Eulemberger K, Elze K, Scifert S. Beiträge zum Blutbild der in Zoologischen Gärten gehaltenen Grosskatzen (Pantherini) 2. Mitteilung: Zum Blutbild klinisch gesunder

Sibirischer tiger (*Panthera tigris altaica*). [Contributions to hematology of large felines (*Pantherini*) in zoos 2. Hematology of clinically healthy Siberian Tiger (*Panthera tigris altaica*)]. Zoological Garten. 1977;**6**:393-412

[12] Kaoud HA. Veterinary Cardiology. 1st ed. UK: CreateSpace; 2017. ISBN: 9781543167122

[13] Kaoud HA. Cardiovascular and Endocrine Disorders in Pets. 1st ed. UK: CreateSpace; 2017. ISBN: 978-1543210422

Chapter 3

Conservation Physiology of Tigers in Zoos: Integrating Stress Physiology and Behaviour to Monitor Their Health and Welfare

Edward Narayan, Nagarajan Baskaran and Janice Vaz

Additional information is available at the end of the chapter

http://dx.doi.org/10.5772/intechopen.69780

Abstract

Big cats in zoos can face challenges associated with captive environments such as inadequate biological adaptation, increased occurrence abnormal behaviour and health-related problems. Conservation physiology is an emerging theme and a dynamic field of research, which aims to reduce these challenges of big cats captive management programmes through new scientific research integrating physiology and behaviour. This field of research applies cutting-edge physiological tools (e.g. non-invasive reproductive and stress hormone monitoring) in combination with traditional methods of behaviour and veterinary health assessments to provide a holistic account of how big cats respond to the captive environment. This book chapter discusses the applications of conservation physiology tools in the captive management of tigers in zoos. Our goal is to bolster tiger captive management in zoos by studying their stress physiology. Overall, the application of conservation physiology tools into captive management programmes for tigers and other big cat species can provide valuable information for evaluating and managing stress, thus improving tiger welfare.

Keywords: conservation physiology, stress endocrinology, zoo biology, animal welfare, behaviour, human-animal interactions, breeding, management

1. Introduction

The world's largest feline species, the tiger, is on the brink of extinction. Tigers are globally listed as 'Endangered' on the International Union for Conservation of Nature (IUCN) Red List of Threatened Species [1]. Captive breeding programmes in zoos around the world have been established to create new home for tigers while also giving a unique opportunity for people and researchers to observe and study these majestic species from surrounds of large closed or open living enclosures. Tiger poaching and hunting has been happening from centuries long ago as ancestral humans became fascinated with their skin and body parts (e.g. tooth) as trophies and souvenirs. Humans are the single most serious threat for the survival of the species. Increasing human population around native tiger habitats (e.g. Southeast Asia) in addition to land clearing and the tiger territorial behaviour (the animal requires access to large habitats to source prey) have led to human-tiger conflicts [2]. In fact, tigers have lost over 90% of their native habitat within the last century, and the currently extant six tiger sub-species are all listed as endangered by the IUCN. Apart from the broader anthropogenic-induced factors that have caused the declines of wild tiger populations such as climate change, land clearing and poaching by humans, the proximate factors, such as diseases and health-related problems, should also be considered as major contributing factors that impact on the survival of tigers [3].

Captive breeding has been a key component of the Tiger Global Conservation Strategy (GCS), which was conceptualised during the international tiger workshop in 1992 at the Edinburgh Zoo, Scotland. The Tiger GCS is a plan for managing tigers internationally, linking captive breeding programmes (ex situ conservation) with in situ conservation [4]. The Tiger GCS supports the long-term management of captive as well as wild tiger populations. Captive populations of tigers should serve as genetic and demographic reservoirs, which will support the wild tiger recovery programmes. The Tiger GCS aims to identify research priorities and technologies that can be transferred between captive and wild conservation efforts. Captive breeding programme for tigers is one of the primary goals for the Tiger GCS, and long-term data on the health and welfare of tigers housed at individual zoos will be valuable resource for identifying the challenges and coming up with strategies to improve the success of tiger captive breeding.

Captive environment can provide safe havens for animals if appropriate environmental enrichment is available so that animals can adequately perform basic life history traits (e.g. foraging, territoriality, social behaviours, resting, mating and nursing young ones). In the absence of appropriate environmental enrichment (e.g. natural and complex enclosures), tigers tend to develop behavioural stereotypes, such as pacing, self-mutilation, aggressiveness, loss of appetite and increased reproductive failure [5]. There are evidences to show that animal-related factors such as sub-species, sex and age also can relate to differences in tiger behaviour and environmental need in captivity [6]. Therefore, the ultimate challenge with captive breeding programmes for large mammals like tigers is to manage the delicate balance between the species' psyche (psychological and physiological states), the phenotype (behaviour) and the environment (e.g. enclosure features, nutrition, human interactions). For example, scientific knowledge on the physiological reaction of the tigers to novel environments will be valuable for our better understanding of this majestic cat species' requirements for biological adaptation in captivity.

This chapter is based on the emerging and valuable science discipline of conservation physiology, which integrates the traditional fields of animal conservation ecology, behaviour, genetics, nutrition and species ethology to apply innovative physiological tools that can provide new knowledge on the species' biology [7]. The primary goal of conservation physiology is to support the on-ground conservation and management efforts through integration of physiological tools into target species conservation programmes. Conservation physiology research has been applied successfully to study the stress physiology and reproductive biology of numerous majestic and iconic animal species, and the technologies (e.g. non-invasive reproductive and stress hormone monitoring) have been immensely beneficial for wildlife captive breeding programmes and conservation translocations [7].

2. Stress physiology of tigers

Most wildlife animals like big cats perceive their environment or surroundings using a combination of behavioural and physical adaptations, such as excellent swimming capabilities, caring for young ones and skin coloration, which blends well into the grassland. Stress is an inbuilt component of the tiger's life history, and stress is not inherently bad because it prepares the animal for behavioural responses through energy mobilising processes, e.g. active hunting, mating, foraging and so forth. Glucocorticoids or stress hormones (e.g. corticosterone and cortisol in tigers) are released into systemic blood circulation during the initiation of the physiological stress response [8]. This occurs when a stimulus (e.g. ungulate prey source) is visualised by the tiger and the neuroendocrine signals in the brain are received by complex set of neurons in the hypothalamic brain region, the paraventricular nucleus (PVN). The PVN is responsible for the initiation of the physiological stress response through activation of the

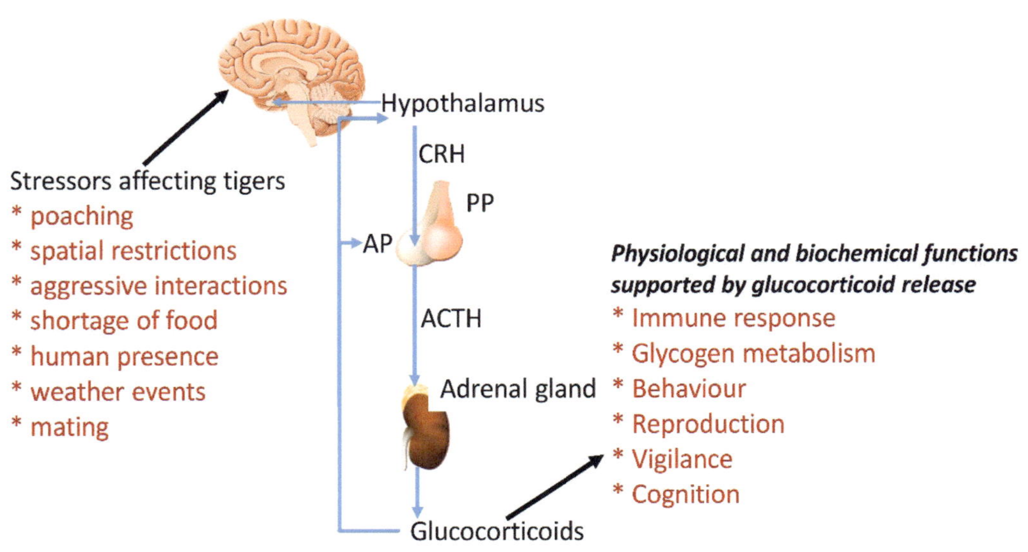

Figure 1. Pathway diagram showing the hypothalamus-pituitary-adrenal axis (HPA axis) and the feedback regulation of glucocorticoid hormone secretion by the HPA axis. CRH, corticotropin-releasing hormone; AP, anterior pituitary; PP, posterior pituitary; ACTH, adrenocorticotropic hormone.

hypothalamus-pituitary-adrenal (HPA) axis. Neural signals from the PVN lead to the secretion of corticotrophin-releasing hormone (CRH) from the hypothalamus and are released at the median eminence to the capillary beds of the hypothalamic-hypophyseal portal system. CRH stimulates the release of adrenocorticotropic hormone (ACTH) from the anterior pituitary, and ACTH ultimately stimulates the adrenal corticotrophes to secrete glucocorticoids (GCs). GCs are responsible for changing metabolic status of the animal (intermediary metabolism), as well as other key physiological processes, such as immune response, skeletal growth, cardiovascular function, reproduction and cognition. GCs perform physiological and biochemical actions in various cells in the body through the help of GC receptors. To maintain homeostasis, the GCs also regulate the higher brain centres, the hypothalamus and the anterior pituitary to stop CRH and ACTH release, thus enabling the levels of systemic GCs to return to basal levels once the stressor has been removed (e.g. few hours after the kill) (see **Figure 1**).

3. Tiger stress biology research in captivity

3.1. Conservation physiology approach: integrating stress biology, environmental factors and welfare

Capture and captivity of tigers can lead to physiological stress impacting on their wellbeing. Measurement of stress in tigers during human interventions such as capture is necessary to quantify potential impact of the external stimulus on the animal HPA axis. Stress could lead to negative consequences such as increased stereotypic behaviour, reproductive infertility and recurring health issues. Animals display a suit of behaviour in response to their environment to show emotional connectedness with their environment. These behaviours may either be signs of healthy animal or an indicator of poor welfare. Therefore, behaviour can be considered as the animal's first line of defence in response to environmental change [9]. Stereotypic or repetitive behaviour in animals serves as indicators or stress, usually associated with elevated glucocorticoid levels [10]. Some of the stressors commonly faced in the captive environment include elements of the enclosure [11], certain aspects of animal care routines [8] and construction noise [12]. Assessment of the welfare of tigers in captivity will be more comprehensive through the integration of behavioural and physiological measures [13].

The captive environment has multiple extrinsic factors, such as duration and nature of light, sound, odours or temperatures, over which the tigers have little control [14]. Optimal psychological and physiological wellbeing of captive care will require the identification and provision of environmental conditions that can promote naturalistic behaviour and adaptation of tigers in captivity [15]. This is defined in animal husbandry principle as environmental enrichment. Therefore, by understanding the stress biology, performance (health and wellbeing) and environmental needs of tigers, zoo managers can identify why some individuals tend to perform poorly or show behavioural signs of stress more often than others. This integration of animal biology (e.g. stress physiology and reproductive behaviour) with environmental

http://dx.doi.org/10.5772/intechopen.69780

factors and monitoring of fitness directly highlights the concept of conservation physiology (see **Figure 2**).

Through the applications of conservation physiology tools, such as non-invasive reproductive and stress hormone monitoring, quantitative data related to tiger biology can be obtained and can be related to the environmental conditions available to them and their overall health and wellbeing. Zoos urgently need to apply these conservation physiology tools into animal welfare and health assessment programmes.

There has been a handful of research based on the stress biology of tigers in zoos, applying the conservation physiology framework (**Figure 2**).

Sajjad et al. [16] studied the effect of captivity on the blood plasma cortisol level and behavioural pattern of tigers managed in two local zoos (indoor versus outdoor enclosures) in Pakistan. The results showed no significant difference in plasma cortisol levels between the two enclosure types; however, behavioural analysis showed differences in pacing percentage between the wildlife park and the zoo (time spent in stereotyping was higher for tigers living in closed enclosures). Plasma cortisol levels were higher when blood collection was done using squeeze caging of the studied tigers.

In a recent research, Vaz et al. [17] applied the concept of conservation physiology to evaluate the stress biology and stereotypic behaviour of tigers (41 Royal Bengal tigers) across six zoos in India. We used non-invasive stress hormone monitoring using faecal corticosterone enzyme immunoassays to quantify the levels of stress in the studied tigers. Our research has found significant correlation between stereotypic behaviour and stress levels in healthy tigers. Furthermore, the stereotypic extent was reduced with environmental enrichment (e.g. increased enclosure size, presence of pools and stones, social interactions with conspecifics

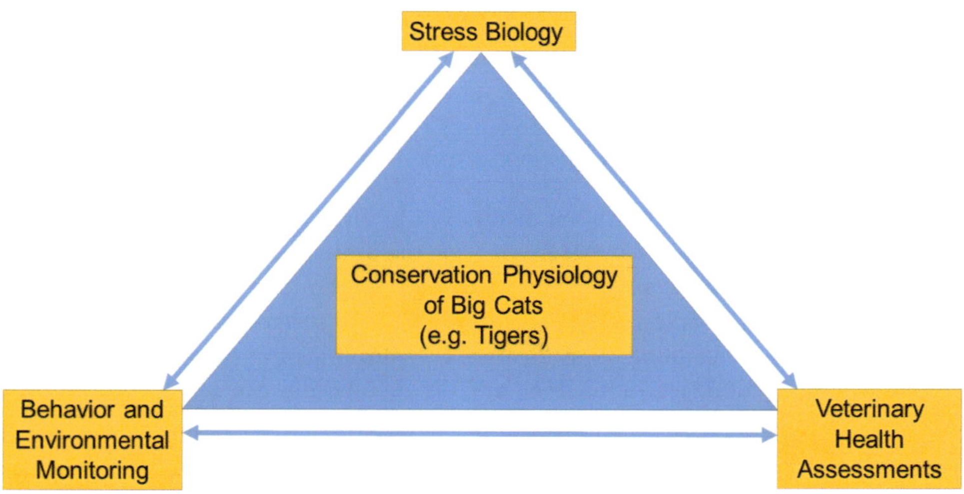

Figure 2. Conservation physiology conceptual framework that interconnects key components of research, including stress biology, behaviour and environmental monitoring and tiger veterinary health assessments.

and positive keeper interactions with the tigers). Stress levels indicated chronic stress in tigers with health problems.

Furthermore, Narayan et al. [8] used faecal cortisol monitoring to investigate the stress physiology of two tiger sub-species, the Bengal and Sumatran tigers, held in two Australian zoos. This study demonstrated that tiger stress levels were significantly different between sexes and zoos, matching with previous studies that also found higher average stress levels in females than males for other felids [18]. Narayan et al.'s [8] study also showed that individual tigers responded differently to the stressor type, intensity and duration. This highlights the phenomenon of unique personalities or coping abilities in animals [19]. Our 2013 study [8] showed that tigers participating in routine activities, such as visitor interactions, expressed higher levels of stress than those individuals that did not participate in these activities. However, tigers from another local Australian zoo showed lower levels of stress when they participated in display activities. These results highlighted that tigers have individual differences in their physiological stress responses, and it also shows that research should quantify stress biology as a quick way of determining the physiological health of the tigers. Variation in the physiological stress responses will indicate the physiological resilience or potential signs of sublethal stress in individual tigers. Non-invasive hormone monitoring provides an early window of opportunity for researchers to see into the stress biology of these majestic feline species.

Dembiec et al. [20] studied the effects of transportation on tigers using combination of stress biology measurements (faecal cortisol and respiration rate) and behaviour during and after transport (activity level, pacing behaviour, investigative behaviour and ear position). Both respiration rate and cortisol profiles showed significant changes after transport (respiration rate increased from 94.6 to 132.3 breaths/10 min after release into enclosures), while faecal cortisol levels remained elevated for up to 12 days in some tigers. Interestingly, the researchers also discovered that 'naïve tigers' showed higher incidence of pacing and higher faecal cortisol than conspecifics that were habituated to the procedure. The data highlighted requirement for appropriate enrichment training and controlled exposure to novel environments in captive tigers.

3.2. Key considerations for non-invasive endocrine monitoring in tigers

Traditional method of stress hormone evaluation in animals included blood collection; however, the process itself can be strenuous for the animal, and the pulsatile nature of systemic glucocorticoid concentrations makes interpretation very difficult. The dynamic field of conservation physiology research has vigorously tested and validated non-invasive methods of stress hormone quantification in wildlife, using excreta samples instead of blood. GCs are metabolised in the liver, and GC metabolites can be found in urine and/or faeces. Faecal-based GC metabolite testing has been used for felids, including tigers [20]. Faecal samples can be collected routinely as part of captive husbandry in zoos. Faecal glucocorticoid metabolites provide a pooled level of GCs that have already participated in the physiological stress response and ready for quantification through excreta analysis. Thus, faecal-based GC monitoring can be used to obtain longitudinal profiles of stress levels in captive animals [21]. One of the key considerations when non-invasive hormone monitoring is being attempted in a species (e.g. tigers) is to obtain data related to time lag of GCs between the activation of the physiological stress response and the collection of GC metabolites in excreta. Typically, hormonal challenge using ACTH is conducted with faecal collections for several days before and after the exogenous hormonal challenge. The peak response of FGMs is then used to identify the time lag, which is valuable information for designing the study. In the absence of ACTH challenge, other stressors such as pregnancy, veterinary anaesthesia can be used routinely to validate faecal-based hormone monitoring tools. Earlier, Narayan et al. [8] validated faecal-based cortisol monitoring in zoo tigers using veterinary check (blood collection) as a physical stressor. Narayan et al. [8] conducted biological validation by collecting faeces 5 days before and 5 days after blood was taken from four male and five female Bengal tigers from a local zoo in Australia. The results showed that the mean levels of faecal cortisol increased by 138 and 285% in the male and female tigers within 1 day after bloods were taken, returning to baseline in 5 days.

In earlier research, Powell et al. [12] validated faecal cortisol monitoring in the Siberian tiger (*Panthera tigris altaica*) using ACTH challenge, demonstrating 2 days time lag of faecal cortisol metabolites in tiger faecal extract after ACTH challenge. The other key considerations of designing comprehensive research in tiger conservation physiology are to consider the effect of natural environmental conditions on the decay of GC metabolites in faecal samples and also to consider the age of samples before analysis. Recently, research from my lab group has demonstrated that faecal cortisol metabolites are stable up to 2 days and homogenising fresh faecal pellets will help reduce the variation in faecal GC metabolite results when small sample sizes are used [22].

Furthermore, levels of GC metabolites in excreta can also be influenced by a suit of internal and external factors. It is well documented in wildlife studies that GCs are metabolic hormones, and the energy mobilisation hypothesis states that GC levels will be higher during more energy-demanding periods, such as breeding, foraging, migration and active search for food and substrate prior to hibernation [23]. Millspaugh and Washburn [24] reviewed the factors that can influence GC levels, such as season, food and habitat availability, reproductive variability, age and gender. Therefore, future research should combine intrinsic (animal related) and extrinsic (environment related) factors into stress hormone monitoring research to derive biologically meaningful interpretation and meaning of the hormonal data.

Furthermore, GC concentrations have been suggested to be significantly affected by reproductive status or cyclic phase of females in some species (e.g. carnivores [25]), which could be explained by increased metabolic demands associated with reproduction [26]. For example, Goymann et al. [25] found that GC levels in lactating female spotted hyenas (*Crocuta crocuta*) were higher than levels for non-lactating females in the wild. Natural variation in GCs tends to complicate the analysis and interpretation of large datasets; however, if researchers are able to obtain as much information on the biology and environmental factors surrounding their study subject, then appropriate statistical models can be applied to test the significance of GC data. Non-invasive reproductive and stress hormone monitoring has been successfully used to monitor tigers in managed habitats and zoos [27, 28].

4. Conclusions

Tiger conservation physiology is a useful new research theme that can significantly bolster captive breeding programmes for the species. Through the applications of non-invasive hormone monitoring tools into tiger health and welfare assessments, more detailed information can be obtained related to their physiological coping capacity towards the captive environment. Zoos can actively collaborate with university-based researchers and apply research students to develop new projects based on tiger conservation physiology. Conservation physiology research has plenty to offer to tiger biology and welfare research and now is the time to apply non-invasive physiological tools into zoo programmes, so that with access to new data on tiger conservation physiology, this majestic cat can be better managed and protected in zoos.

Acknowledgements

The authors are highly thankful to senior mentors, students and collaborators who have worked on the tiger conservation physiology research discussed in this chapter. We highlight no conflict of interest. Dr. Edward Narayan conceptualised the research chapter and wrote the chapter. Janice Vaz conducted master's research based on the conservation physiology of tigers in Indian Zoos supervised by Drs. Narayan and Baskaran.

Author details

Edward Narayan[1]*, Nagarajan Baskaran[2] and Janice Vaz[2]

*Address all correspondence to: e.narayan@westernsydney.edu.au

1 Western Sydney University, School of Science and Health, Locked Bag 1797, Penrith, NSW, Australia

2 A.V.C. College (Autonomous), Mannampandal, India

References

[1] Goodrich J, Lynam A, Miquelle D, Wibisono H, Kawanishi K, Pattanavibool A, Htun S, Tempa T, Karki J, Jhala Y, Karanth U. *Panthera tigris*. The IUCN Red List of Threatened Species 2015: e.T15955A50659951. Available from: http://dx.doi.org/10.2305/IUCN.UK.2015-2.RLTS.T15955A50659951.en. [Accessed: 09-04-2017]

[2] Nyhus P, Sumianto, Tilson R. The tiger-human dimension in southeast Sumatra. In: Seidensticker J, Christie S, Jackson P, editors. Riding the Tiger: Tiger Conservation in Human-Dominated Landscapes. Cambridge: Cambridge University Press; 1999. pp. 144-145. ISBN 0-521-64835-1

[3] Shrivastav AB. Wildlife health: A new discipline: Essential for tiger conservation. INTAS –POLIVET. 2001;**2**:134-136

[4] Tilson R, Brady G, Traylor-Holzer K, Armstrong D, editors. Management and Conservation of Captive Tigers. 2nd ed. MN: Apple Valley: Minnesota Zoo; 1994. p. 136

[5] Breton G, Barrot S. Influence of enclosure size on the distances covered and paced by captive tigers (*Panthera tigris*). Applied Animal Behaviour Science. 2014;**154**:66-75. DOI: 10.1016/j.applanim.2014.02.007

[6] Pitsko L. Wild tigers in captivity: A study of the effects of the captive environment on tiger behavior. [Masters Thesis]. Virginia Polytechnic Institute and State University; 2003

[7] Narayan E. Non-invasive reproductive and stress endocrinology in amphibian conservation physiology. Conservation Physiology. 2013;**1**:1-16. DOI: 10.1093/conphys/cot011

[8] Narayan E, Clark G, Martin-Vegue P, Parnell T, Mucci A, Hero J-M. Faecal cortisol metabolite levels in Bengal (*Panthera tigris tigris*) and Sumatran tigers (*Panthera tigris sumatrae*). General and Comparative Endocrinology. 2013;**194**:318-325

[9] Mench, JA. Why it is important to understand animal behavior. ILAR Journal. 1998;**39**:20-26. DOI: 10.1093/ilar.39.1.20

[10] Carlstead K. Effects of captivity on the behavior of wild mammals. In: Kleiman DG, Allen ME, Thompson KV, Lumpkin S, editors. Wild Mammals in Captivity—Principles and Techniques. Chicago, London: The University of Chicago Press; 1996. pp. 317-333

[11] Bashaw JH, Kelling AS, Bloomsmith MA, Maple TL. Environmental effects on the behavior of zoo-housed lions and tigers with a case study of the effects of a visual barrier on pacing. Journal of Applied Animal Welfare Science. 2007;**10**:95-109. DOI: 10.1080/10888700701313116

[12] Powell DM, Carlstead K, Tarou LR, Brown JL, Monfort SL. Effects of construction noise on behavior and cortisol levels in a pair of captive giant pandas (*Ailuropoda melanoleuca*). Zoo Biology. 2006;**25**:391-408. DOI: 10.1002/zoo.20098

[13] Wielebnowski NC, Fletchall N, Carlstead K, Busso JM, Brown JL. Noninvasive assessment of adrenal activity associated with husbandry and behavioural factors in the North American clouded leopard population. Zoo Biology. 2002;**21**:77-98. DOI: 10.1002/zoo.10005

[14] Morgan KN, Tromborg CT. Sources of stress in captivity. Applied Animal Behaviour Science. 2007;**102**:262-302. DOI: 10.1016/j.applanim.2006.05.032

[15] Shepherdson DJ, Mellen JD, Hutchins M, editors. Second Nature: Environmental Enrichment for Captive Animals. Washington, DC: Smithsonian Institution Press; 1998. p. 350

[16] Sajjad S, Farooq U, Anwar M, Khurshid A, Bukhari SA. Effect of captive environment on plasma cortisol level and behavioral pattern of Bengal tigers (*Panthera tigris tigris*). Pakistan Veterinary Journal. 2011;**31**:195-198

[17] Vaz J, Narayan EJ, Kumar RD, Thenmozhi K, Thiyagesan K, Baskaran N. Comparative evaluation of stereotypical behaviours and physiological stress in tigers and leopards from six Indian zoos. PLoS One. 2016;**12**(4):e0174711

[18] Brown JL, Wildt DE. Assessing reproductive status in wild felids by noninvasive faecal steroid monitoring. International Zoo Yearbook. 1997;**35**:173-191. DOI: 10.1111/j.1748-1090.1997.tb01208.x

[19] Gosling SD, John OP. Personality dimensions in non-human animals: A cross-species review. Current Directions in Psychological Science. 1999;**8**(3):69-75. DOI: 10.1111/1467-8721.00017

[20] Dembiec DP, Snider RJ, Zanella AJ. The effects of transport stress on tiger physiology and behavior. Zoo Biology. 2004;**23**:335-346. DOI: 10.1002/zoo.20012

[21] Nizeyi CN, Monfort SL, Taha N, Cranfield M, Linda P, Gilardi K. Detecting adrenocortical activity in gorillas: A comparison of faecal glucocorticoid measures using RIA versus EIA. International Journal of Animal and Veterinary Advances. 2011;**3**:104-116

[22] Parnell T, Narayan EJ, Nicolson V, Martin-Vegue P, Mucci A, Hero J-M. Maximizing the reliability of non-invasive endocrine sampling in the tiger (*Panthera tigris*): Environmental decay and intra-sample variation in faecal glucocorticoid metabolites. Conservation Physiology. 2015;**3**:cov053. DOI: 10.1093/conphys/cov053

[23] Naidenko, SV, Ivanov EA, Lukarevskii VS, Hernandez-Blanco JA, Sorokin PA, Litvinov MN, Kotlyar AK, Rozhnov VV. Activity of the hypothalamo–pituitary–adrenals axis in the Siberian tiger (*Panthera tigris altaica*) in captivity and in the wild, and its dynamics throughout the year. Biology Bulletin. 2011;**38**:301-305. DOI: 10.1134/S1062359011030095

[24] Millspaugh JJ, Washburn BE. Use of fecal glucocorticoid metabolite measures in conservation biology research: Considerations for application and interpretation. General and Comparative Endocrinology. 2004;**138**:189-199. DOI: 10.1016/j.ygcen.2004.07.002

[25] Goymann W, East ML, Wachter B, Höner OP, Möstl E, Van't Holf TJ, Hofer H. Social, state-dependent and environmental modulation of fecal corticosteroid levels in free-ranging female spotted hyenas. Proceedings of the Royal Society: Biological Sciences. 2001;**268**:2453-2459. DOI: 10.1098/rspb.2001.1828

[26] Touma C, Palme R. Measuring fecal glucocorticoid metabolites in mammals and birds: The importance of validation. Annals of the New York Academy of Sciences. 2005;**1046**:54-74. DOI: 10.1196/annals.1343.006

[27] Bhattacharjee S, Kumar V, Chandrasekhar M, Malviya M, Ganswindt A, Ramesh K, et al. Glucocorticoid stress responses of reintroduced tigers in relation to anthropogenic disturbance in Sariska Tiger Reserve in India. PLoS One. 2015;**10**:e0127626. DOI: 10.1371/journal.pone.0127626

[28] Umapathy G, Kumar V, Wasimuddin, Kabra M, Shivaji S. Detection of pregnancy and fertility status in big cats using an enzyme immunoassay based on 5a-pregnan-3a-ol-20-one. General and Comparative Endocrinology. 2013;**180**:33-38. DOI: 10.1016/j.ygcen.2012.10.009

Section 2

Ecology and Evolution

Chapter 4

Applied Biotechnologies in the Conservation of Wild Felids: *In vitro* Embryo Production and Cellular Regenerative Therapies

Daniel Veraguas, Diana Echeverry,
Fidel Ovidio Castro and Lleretny Rodriguez-Alvarez

Additional information is available at the end of the chapter

http://dx.doi.org/10.5772/intechopen.71311

Abstract

This review consist of a complete recompilation of advances made in *in vitro* embryo production (IVP) and cellular therapies in the domestic cat and wild felids species. Actually, the domestic cat is considered a valuable model for the development of assisted reproductive techniques and cellular regenerative therapies that might be used in the conservation of endangered felids. The *in vitro* embryo production technologies such as *in vitro* fertilization (IVF), intracytoplasmic sperm injection (ICSI) and somatic cell nuclear transfer (SCNT) have been applied in different species of small wild felid and big felids, resulting in live birth in some cases. However, more studies are needed to improve the efficiency of these techniques and maximize their use in the reproduction and conservation of wild felids. On the other hand, the mesenchymal stem cells (MSCs) therapies have been increasingly used in the veterinary medicine. Recent studies had reported the use of MSCs in the treatment of some chronic diseases in the domestic cat. For these reasons, the MSCs therapies are projected as a promising alternative for the treatments of chronic diseases that might affect wild felids that live in zoos, which could be caused by their captive environment conditions.

Keywords: wild felids, domestic cat, *in vitro* fertilization, somatic cell nuclear transfer, mesenchymal stem cells

1. Introduction

Actually, the domestic cat is considered a valuable model for the development of assisted reproductive technologies (ARTs) and cellular regenerative therapies that might be used in the conservation of endangered felids [1, 2]. At present, according to the Red List of Endangered Species of the IUCN, 17 species of wild felid are classified as endangered and 8 as near threatened [3]. This has led to the implementation of programs based in ARTs for the genetic preservation of endangered felids in zoological institutions from the USA and countries of Latin America like Mexico and Brazil [4, 5]. The *in vitro* embryo production (IVP) systems are a valuable tool that allow obtaining information regarding the developmental biology and might help the genetic preservation of endangered felids. The *in vitro* fertilization (IVF) is the most used IVP technique in the domestic cat and it has been used to produced embryos from several wild felid species [2, 6, 7]. The intracytoplasmic sperm injection (ICSI) has been used in the domestic cat with successful live birth [8, 9]. In some wild felid species, the ICSI has been used as a tool to help the embryo production in the cases of low sperm motility [8, 10]. The somatic cell nuclear transfer (SCNT) is a complex technique with a reduced efficiency compared to the techniques previously described. However, the SCNT allows the generation of genetically identical individuals and might help to preserve animals with a high genetic value [11]; the first domestic cat generated by SCNT was born in 2002 [12]. Since then, several research groups have tried to replicate the same achievement in wild felids. In 2003, the first African wildcat (*Felis silvestris lybica*) was born by interspecific somatic cell nuclear transfer (iSCNT), which proved the efficacy of this technique for the conservation of endangered felids [13, 14]. In general, the IVP techniques in felids still have low efficiency compared to other species [15]. For this reason, more studies are needed to improve these problems.

Regarding cellular therapies, cellular resources that might be potentially used in cell therapy, have been obtained in different species. The use of methodologies for cellular reprogramming has allowed the production of pluripotent and multipotent cells. However, it has been preferred that those methods that do not implicate a dependence on the expression of exogenous genes or changes at the cellular level may affect the viability of the cell for the use in therapy.

The use of stem cell for regenerative therapy in cats has been stimulated by the possibility to treat severe diseases such as chronic renal failure that is the main cause of death in felines, asthma and chronic feline gingivostomatitis. The domestic cats also represent a good model for the study of the efficacy of stem cell therapies for human diseases. Furthermore, it has been postulated that the use of a less differentiated cell type, as the mesenchymal stem cells, like nucleus donor, can improve the efficiency of the SCNT technique, which could be applicable in the conservation of endangered felid species.

2. *In vitro* embryo production systems in the domestic cat and wild felids

The studies related to the use of ARTs in felids have increased considerably during the last years [16]. In 1990s, several studies reported the live birth of different big felid as the cheetah (*Acinonyx jubatus*), Siberian tiger (*Panthera tigris altaica*), clouded leopard (*Neofelis nebulosa*) and the Puma (*Puma concolor*) by artificial insemination (AI) [17–20]. However, scares information has been published regarding the *in vitro* embryo production in big felids. On the other hand, several studies of IVF, ICSI and SCNT have been made in different species of small wild felids [16]. The *in vitro* embryo production systems are a valuable tool that allows the study of the gamete interaction, the early embryonic development and the rescue of genetic material. Based on gathered information, it seems that the improvement of IVP technologies in felids might have a significant impact to preserve the endangered big felids.

2.1. *In vitro* fertilization (IVF)

The first IVF report in felids was made in the domestic cat in 1970, in which cat oocytes were co-cultured *in vitro* with spermatozoa that were previously incubated in the uterus [21]. Subsequently, Goodrowe et al. described the first live offspring of domestic cat by IVF after the transfer of embryos at 2–4 cell stage [22]. Actually, the IVF is the most used technique for the *in vitro* embryo production in the domestic cat and several wild felids species [2].

2.1.1. *In vitro fertilization in the domestic cat*

A series of studies have been made to improve the efficiency of the IVF. However, the domestic cat embryos generated by IVF have a reduced developmental capacity compared to the *in vivo* produced embryos [23]. In the domestic cat, the percentage of blastocysts produced from oocytes matured and fertilized *in vitro* is low compared to those matured and fertilized *in vivo* (10–50 vs. 50–66%, respectively) [15]. The domestic cat embryos generated *in vitro* suffer a developmental block at the morula stage, which is attributable to deficient conditions in the oocyte *in vitro* maturation and in the *in vitro* embryo culture [24, 25]. However, some research groups have overcome this inconvenience, reporting blastocyst rates of 50–70% approximately after 6, 7 or 8 days of culture [2, 16, 26–28] (**Table 1**).

One of the biggest problem of the IVP systems in the domestic cat, is the low competence of the *in vitro* matured oocytes. For this reason, some research groups have decided to use gonadotrophins to induce the *in vivo* maturation of cat oocytes. Pope have used porcine FSH (pFSH), to induce follicular recruitment and porcine LH, (pLH) to induce oocyte maturation, which have allowed the successful embryo generation by IVF, ICSI and SCNT [16]. Likewise, Yu et al. have described the use of eCG and hCG for the *in vivo* maturation of cat oocytes and their subsequently use in parthenogenetic activation, IVF and SCNT [29].

Our research group evaluated the individual effect of both gonadotrophins (pFSH and eCG) in the morphological quality and gene expression pattern of cumulus-oocyte complexes (COCs). Regarding the pFSH treatment, our study demonstrated that pFSH enhanced the morphological quality of COCs and increased the expression of the gonadotrophin receptor *LHCGR*. Furthermore, these results were related to an improved *in vitro* embryo development with an increased blastocyst and hatching blastocyst rates compared to the control untreated group (30.5 vs. 13.1% and 13.2 vs. 1%, respectively) [30]. Likewise, the treatment of domestic cat with eCG enhanced the morphological quality and the expression pattern of the COCs. The eCG treatment increased the relative expression of gonadotrophin receptors (*FSHR* and *LHCGR*) and gonadotrophin-induced genes (*EGFR*, *EGR1* and *ESR2*) in the COCs, which might be related to an enhanced oocyte competence [31]. Subsequently, we observed that the oocytes recovered from cats treated with eCG had an enhanced developmental capacity after

Oocyte source	*In vitro* embryo production by IVF					Reference
	Cleavage rate (%)	Morula rate (%)	Blastocyst rate (%)	Pregnancy rate (%)	Live birth rate (%)	
In vivo maturation	62.9–88.2	40–47	33.2–71.3	31–83.3	31–83.3	[2, 8, 22, 26, 29, 34]
In vitro maturation	50.2–71.5	52.2–62.7	30.0–61.1	40	20	[2, 27, 28, 105]
eCG + IVM	64.9	74.2	32.9	66.6	33.3	[Unpublished data]

The ranges showed were derived from several results of the respective studies cited.

- Pregnancy rate was derived from the total pregnancies established/total transferred recipients, according to each reference.
- Live birth rate correspond to the transferred recipient that delivered healthy kittens/total transferred recipient, according to each reference.

Table 1. *In vitro* and *in vivo* development of domestic cat embryos after IVF using oocytes from different sources.

parthenogenetic activation, which was reflected in a higher blastocyst rate compared to the control untreated group (32.8 and 16.9%, respectively) [unpublished data]. No differences on oocyte competence were observed between pFSH and eCG treatments. For this motive, we choose the eCG treatment for the ovarian stimulation in domestic cats. The eCG has a longer half-life than pFSH and only a unique dose is required, which results in a more practical protocol and less stress for the animals.

Actually, our research group has used the mild ovarian stimulation with eCG for the *in vitro* embryo production in the domestic cat. The domestic cat was treated only with a unique dose of 200 IU of eCG and then the recovered COCs were *in vitro* matured. We choose this protocol because no additional improvement was observed in the developmental competence using *in vivo* matured oocytes after eCG and hCG treatment. After parthenogenetic activation, no differences were observed in the blastocyst rate between the eCG and IVM group (32.8%) and the eCG and hCG group (31.4%) [unpublished data].

This protocol was used in the production of embryos by IVF (**Table 1**). Additionally, we evaluated the *in vivo* development of the generated blastocyst. Three embryo transfers were made, 23 day-7 blastocysts were transferred to two domestic cat (15 and 13 blastocysts, respectively). In both cats pregnancy was established, but only the cat that received 15 blastocysts gave birth to one female kitty after 64 days of gestation. Additionally, 8 day-8 blastocysts were transferred to one domestic cat but no pregnancy was established (**Table 1**).

In general, the transfer of felid embryos has been successfully made at earlier stages, at the 1–2 cell stage [14] and morula or early blastocyst stages [16, 32]. Kanda et al. evaluated the transfer of cat embryos at the morula stage, early blastocyst stage after 4–6 days of IVC and blastocyst stage after 7 days of IVC. All the recipients that received embryos at the morula stage (4/4, 100%) and three recipients that received early blastocyst (3/5, 60%) became pregnant.

However, no pregnancies were established after the transfer of 7-days blastocysts (0/3) [33]. It has been postulated that this might be due to the zona pellucida integrity of 7-days cat blastocysts, most of the blastocysts begin to hatch the zona pellucida after day 6 and this might affect negative *in vivo* development [16, 33].

2.1.2. *In vitro fertilization in small wild felids*

The first live birth report was made in the Indian desert cat (*Felis silvestris ornata*). After IVF, 4 Indian desert cat morulae and 10 domestic cat morulae were transferred to the uterine horn of a domestic cat recipient, which gave birth to one live Indian desert cat [6, 34]. This demonstrated that the domestic cat can be successfully used as recipient for small wild felid embryos. The same research group reported later, the live birth of an African wildcat generated by IVF; the embryos were cryopreserved and subsequently transferred to a domestic cat recipient [35]. In subsequent years, the live birth from IVF-derived embryos was reported in the Serval cat (*Leptailurus serval*), Caracal (*Caracal caracal*), Fishing cat (*Prionailurus viverrinus*), Sand cat (*Felis margarita*) and Black-footed cat (*Felis nigripes*) [7, 32, 36, 37]. In these studies, the embryos were transfer at the morula stage after 5 or 6 days of IVC to domestic cat recipients [7] or intra-species recipients [7, 32, 36, 37]. Additionally, the interspecific IVF has been described in the bobcat (*Lynx rufus*) and in the Iberian lynx (*Lynx pardinus*) as a method to prove the fertilization capacity of cryopreserved spermatozoa from these species using domestic cat oocytes [38, 39] (**Table 2**).

2.1.3. *In vitro fertilization in big felids*

Actually, only a few studies have reported the embryo production by IVF in big felid. The first successful report of IVF was made in tiger (*P. tigris*). After IVF, 86 tiger embryos at the 2–4 cell stage were transferred into the oviduct of four recipients. One pregnancy was established and three cubs were delivered 107 days after embryo transfer [40]. Other study demonstrates that the cryopreservation of tiger spermatozoa does not affect the fertilization rate and developmental competence after IVF [41]. Regarding embryo cryopreservation, vitrification is the most efficient method for the cryopreservation of tiger embryos generated by IVF [42]. No more studies of IVF in the tiger have been published.

IVF have been performed in the Puma as well; the semen samples collected from male specimens and used for the IVF showed high degree of pleomorphism (82–99%). Despite that, an overall fertilization rate of 43.5% was obtained [43]. Subsequently, the IVF in the Cheetah was described; after IVF an overall fertilization rate was only 26.6%. However, the fertilization rate varied from 0 to 73.3% among males [44]. Previous reports have described a high proportion of abnormal spermatozoa/ejaculate in the Cheetah (31–97%) [45, 46]. A decade later, it was described the IVF in the lion (*Panthera leo*), being the first report that described the early embryo development in this specie (**Table 2** and **Figure 1**).

In resume, the IVF has been used more in small wild felids than in big felids, this is mainly due to the easier manipulation of small wild felids during the laparoscopic oocyte retrieval and semen collection procedure. Furthermore, in the case of some small felids species, the

Species	*In vitro* embryo production by IVF						Reference
	Fertilization rate (%)	Cleavage rate (%)	Morula rate (%)	Blastocyst rate (%)	Pregnancy rate (%)	Live birth rate (%)	
Indian desert cat	67.0	—	—	—	25.0	25.0	[34]
African wildcat	74.0	—	—	—	66.6	33.3	[2, 35]
Sand cat	—	76.9	—	—	25.0	25.0	[37]
Serval	33.0–68.0	—	—	—	16.6	16.6	[36]
Caracal	—	66.0	—	—	33.0	33.0	[32]
Fishing cat	—	60.0	—	—	9.0	9.0	[32]
Black-footed cat	—	47–70	—	—	40.0	40.0	[7]
Bobcat	46*	—	27*	—	—	—	[38]
Iberian linx	11.5–20.5*	44.7–87.5*	—	—	—	—	[39]
Tiger	63.4	39.0	43.5	30.4	25.0	25.0	[40]
Puma	43.5	—	—	—	—	—	[43]
Cheetah	0–73.3	17.3	—	—	—	—	[44]
Lion	—	53.0	50.0	30.0	—	—	[47]

*Interspecific embryos generated using domestic cat oocytes.

The ranges showed were derived from several results of the respective studies cited.

- Pregnancy rate was derived from the total pregnancies established/total transferred recipients, according to each reference.
- Live birth rate correspond to the transferred recipient that delivered healthy kittens/total transferred recipient, according to each reference.

Table 2. Efficiency of the embryo production by IVF in wild felids.

generated embryos can be transferred to a domestic cat recipient producing live offspring. Additionally, the high pleomorphism incidence in the spermatozoa of some big felid species, reduces the efficient of IVF in those species.

2.2. Intracytoplasmic sperm injection (ICSI)

Actually, the fertilization techniques assisted by micromanipulation are used widely in the human fertility clinics, allowing the *in vitro* embryo production in the cases of low sperm quality or reduced motility [48]. The ICSI is the direct injection of a single spermatozoa into the cytoplasm of a matured oocyte, eliminating the negative effects of low sperm motility [48]. ICSI is projected as a possible alternative for the *in vitro* embryo production in wild felids, where a high degree of morphological abnormality and low motility in the collected sperm have been observed [49].

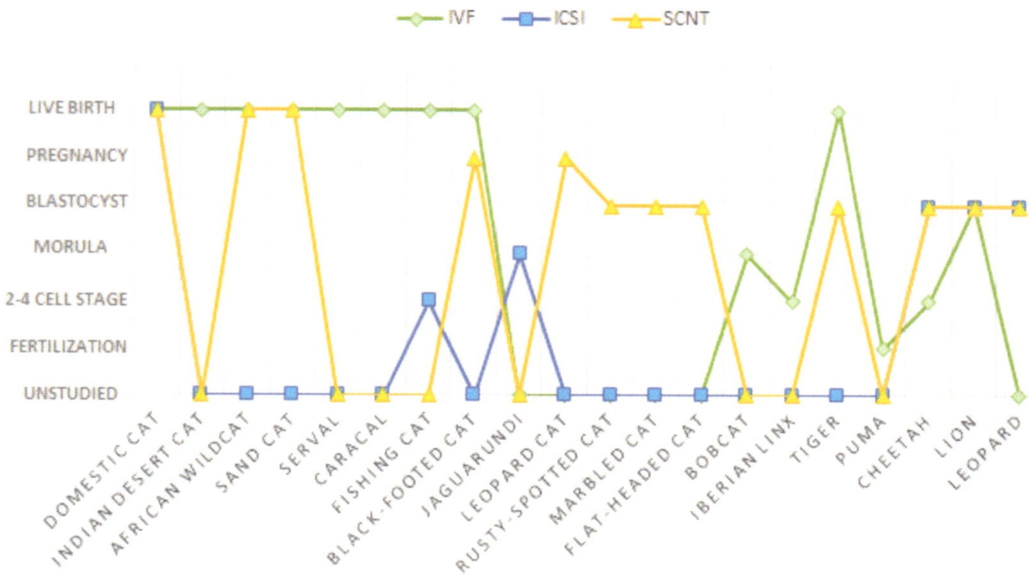

Figure 1. Progresses in the embryo generation by IVF, ICSI and SCNT in the domestic cat and wild felids.

2.2.1. *Intracytoplasmic sperm injection in the domestic cat*

The first ICSI-derived domestic cat embryos were produced using *in vivo* matured oocytes. After the transfer of 10–11 morulae at 5 day of culture into four recipients, two pregnancies were established and three kittens were born [8]. It was described that the embryos generated by ICSI using *in vitro* matured cat oocytes had a decreased developmental competence compared to those derived from *in vivo* matured oocytes [9]. Despite this, the ICSI embryos generated from *in vitro* matured cat oocytes are able to produce live births [9]. Subsequently, the production of cat embryos by ICSI using *in vitro* matured oocytes and frozen epididymal spermatozoa was described. The generated embryos were capable to reach the blastocyst stage [50]. However, the blastocyst rate was low (6.6%) compared to the previously reported by Pope et al. (42.9%) and Gómez et al. (19%) [8, 9, 50].

2.2.2. *Intracytoplasmic sperm injection in small wild felids and big felids*

No many studies related to the ICSI in wild felids have been published. The first report was made in the jaguarundi (*Herpailurus yagouaroundi*). Nine jaguarundi embryos that reached the early morula stage at day 5 of culture were transferred into two domestic cat recipients, but none of them developed to term [8]. Subsequently, it was described the generation of Fishing cat embryos by ICSI, 19 day-2 cleaved embryos were transferred into the oviduct of one fishing cat but no pregnancy was established [32] (**Table 3**).

It has been described that the generation of lion embryos by ICSI using sperm collected by percutaneous epididymal sperm aspiration (PESA) from two vasectomized male lions. After ICSI, all the cleaved embryos reached the morula stage at day 5 of culture [51]. Subsequently,

Species	*In vitro* embryo production by ICSI					Reference
	Cleavage rate (%)	Morula rate (%)	Blastocyst rate (%)	Pregnancy rate (%)	Live birth rate (%)	
Domestic cat	57.0–81.6	50.8–81.0	6.6–42.9	33.3–50.0	22.2–50.0	[8, 9, 50]
Fishing cat	22.0–70	—	—	0	—	[32]
Jaguarundi	55.6	50.0	—	0	—	[8]
Cheetah	66.3–73.6*	—	21.0–32.6*	—	—	[10]
Lion	30.8–60.0	28.6–100	28.6–50	—	—	[51, 52]
Leopard	35.3–73.7*	—	9.8–21.0*	—	—	[10]

*Interspecific embryos generated using domestic cat oocytes.

The ranges showed were derived from several results of the respective studies cited.

- Pregnancy rate was derived from the total pregnancies established/ total transferred recipients, according to each reference.
- Live birth rate correspond to the transferred recipient that delivered healthy kittens/total transferred recipient, according to each reference.

Table 3. Efficiency of the ***in vitro*** embryo production by ICSI in domestic and wild felids.

it was described that the lion oocytes are able to mature *in vitro* and after ICSI the embryos are able to reached blastocyst stage. However, the embryos reached the morula stage at day 6 and the blastocyst stage at day 9, which indicated a slow *in vitro* development [52]. More recently, the interspecific ICSI in the leopard (*Panthera pardus*) and Cheetah using domestic cat oocytes was described. The interspecific embryos were able to develop *in vitro* until the blastocyst stage with a similar rate compared to domestic cat embryos. This demonstrates that the developmental capacity of big felid spermatozoa can be evaluated by interspecific ICSI [10] (**Table 3** and **Figure 1**).

More research is needed to evaluate the ICSI in big felids. The ICSI allows the *in vitro* embryo generation in the cases of poor sperm quality, which is consisted in some big felid species. Furthermore, the interspecific ICSI using domestic cat oocyte offers the possibility to study the *in vitro* developmental capacity in these species.

2.3. Somatic cell nuclear transfer (SCNT)

During the SCNT, the oocyte cytoplasmic factors are capable to reprogram the expression pattern of a somatic cell into a pluripotent embryonic state, allowing it to initiate the early development [53]. The fact that a differentiated cell must be reprogramed to a pluripotent stage made the SCNT a complex technique with a reduced efficiency. Different factors may affect the efficiency of the SCNT, some of these are: the quality of the donor nucleus or the recipient oocyte and the synchronization between the donor nucleus and the oocyte. These factors can lead to an inadequate reprogramming of the donor genome and to a reduced embryo development [11]. Despite this low efficiency, the SCNT is the only technique that may generate genetically identical individuals, which bring the possibility to rescue the genetic material of endangered species.

2.3.1. Somatic cell nuclear transfer in the domestic cat

The first domestic cat produced by SCNT was made using cumulus cells as nucleus donors for metaphase II enucleated oocytes. A live kitten was born 66 days after the embryo transfer [12]. Subsequently, it was described the use of cat fetal fibroblasts and adult fibroblasts as nucleus donors for SCNT. No statistical differences were observed in the fusion, cleavage and blastocyst rates when fetal or adult fibroblasts were used. Furthermore, live birth was obtained using both type of cells [54].

In addition, the advances of the SCNT in felids have led the production of transgenic animals. A domestic cat that expresses the red fluorescent protein was generated using a retroviral vector for the incorporation of the transgene in the somatic cells used as nucleus donor [55]. A year later, it was reported the birth of a transgenic cat that express the green fluorescent protein, which was generated using a lentiviral vector to modify the donor cells [56]. The objective of these studies was to potentiate the use of the domestic cat as a biomedical model for the study of analogues diseases in humans. This proves that these techniques could be implemented to generate genetically identical animals that have integrated coding genes for specific human diseases.

2.3.2. Somatic cell nuclear transfer in small wild felids

The scares availability of gametes is one of the biggest problem for the *in vitro* embryo production in endangered animals due to their reduced population. The interspecific somatic cell nuclear transfer (iSCNT) is proposed as alternative for this problem. It has been demonstrated that the domestic cat oocyte can be used as recipient cytoplast for somatic cells of wild felids, generating embryos of these species [13]. It was reported the live birth of African wildcats by iSCNT using the domestic cat oocyte as recipient cytoplast and a female domestic cat as recipient for the cloned embryos [14]. Subsequently, the same research group described the live birth of three Sand cats by iSCNT [57]. Different studies have tried to replicate those results in other felid species, but no live births have been reported. This may be because the African wildcat and the Sand cat are subspecies closely related to the domestic cat, which might improve the efficiency of cell reprograming during iSCNT. However, it seems when somatic cells from felid species, phylogenetically more distant from the domestic cat are used, the efficiency of the iSCNT tend to decrease.

In the leopard cat (*Prionailurus bengalensis*), the embryo generation by iSCNT was reported, using domestic cat oocytes as cytoplasm recipients. The cloned embryos reached the blastocyst stage. However, after embryo transfer, pregnancy was established but no live births were obtained [58]. Subsequently, the generation of embryos of rusty-spotted cat (*Prionailurus rubiginosus*) and black-footed cat by iSCNT was described. The cloned embryos were capable to develop until the blastocyst stage. After embryo transfer, the rusty-spotted cat embryos were not able to implant. Meanwhile, the transferred black-footed cat embryos were able to establish pregnancy but all the fetuses were reabsorbed later [59]. Similarly, in the marbled cat (*Pardofelis marmorata*) and flat-headed cat (*Prionailurus planiceps)*; after iSCNT, the embryos were able to develop *in vitro* until the blastocyst stage. However, no pregnancies were established after embryo transfer into domestic cat recipients [60].

2.3.3. Somatic cell nuclear transfer in big felids

There are scarce reports regarding to the iSCNT in big felids. The first report was made in the tiger, in which pig oocytes were used as recipient cytoplast. However, a blastocyst rate of only 0.7% was obtained [61]. More recently, leopard, tiger and lion cloned embryos were generated by iSCNT, using rabbit oocytes as cytoplast recipients. This study described that the rabbit oocytes were capable to reprogram big felid somatic cells. However, the blastocyst rate after *in vitro* culture was 5–6% in the three species [62].

On the other hand, several methods have been implemented to improve the low efficiency of the SCNT such as embryo aggregation. This consists of the *in vitro* culture of two or more zona-free embryos together in the well of the well system (WOW) [63]. It was reported that the aggregation of cloned cheetah embryos and cloned tiger embryos generated by iSCNT improves the *in vitro* development in these species. However, the developmental capacity of cheetah and tiger cloned embryos was reduced compared to domestic cat cloned embryos [63, 64] (**Table 4** and **Figure 1**).

In resume, the iSCNT have a reduced efficiency and it seems that these problems are more evident when the species used as nucleus donor are more distant phylogenetically from the domestic cat. This might be related to an incomplete donor nucleus reprogramming. Regarding the big felids, it has been proved that pig and rabbit oocytes are able to reprogram big felid donor cells, but they are not as efficient as domestic cat oocytes. The embryo aggregation has been postulated as a method that improves the developmental capacity of big cat cloned embryos. However, the developmental capacity of these embryos is reduced compared to domestic cat cloned embryos. More studies are needed to improve the reprograming events that allow the development of iSCNT-derived embryos in big felids.

2.4. Gene expression analysis as indicator of developmental capacity in felid embryos

During the early development, the expression of specific genes allows the embryo progress from one stage to the next [65]. For this reason, the relative quantification of mRNA from crucial genes during the early development is considered an adequate indicator of the embryo quality [66]. The pluripotency markers *OCT4*, *SOX2* and *NANOG* and the differentiation markers *CDX2* and *GATA6* play a crucial role during the embryo development in the domestic cat [67]. According to this, an aberrant expression pattern of these genes might be related to a reduced developmental competence of the produced embryos [67].

It has been described that the gonadotrophin treatments in the domestic cat improve the gene expression pattern in the COCs and in the produced embryos. The ovarian stimulation of domestic cat with pFSH enhances the embryo developmental capacity and increases expression of *OCT4* and *GATA6* at the blastocyst stage [30].

Furthermore, in the domestic cat, it has been described that the *in vitro*-produced blastocysts had a decreased relative expression of *OCT4* compared to the *in vivo*-produced blastocysts [67]. This demonstrated that the *in vitro* culture affect negatively the gene expression pattern of domestic cat embryos, which is reflected in their reduced developmental competence.

Species	*In vitro* embryo production by SCNT					Reference
	Clevage rate (%)	Morula rate (%)	Blastocyst rate (%)	Prenancy rate (%)	Live birth rate (%)	
Domestic cat	57.6–98.2	4.0–49.5	2.0–47.7	20.0	20.0	[13, 14, 54, 63]
African wildcat	79.0–89.0	35.0–51.0	17.0–33.0	40.0–50.0	25.0	[13, 14]
Sand cat	83.0–97.0	—	6.0–43.0	30.0–32.0	12.0–25.0	[57]
Black-footed cat	85.2	—	2.6	45.0	0	[59]
Leopard cat	67.4	—	7.3	57.1	0	[58]
Rusty-spotted cat	85.5	—	26.2	0	—	[59]
Marbled cat	93.3	23.3	5	—	—	[60]
Flat-headed cat	96.7	53.3	8.3	0	—	[60]
Tiger	93.7–95.9	18.6–25.4	3.4–12.7	—	—	[64]
Cheetah	87.2–96.7	37.4–38.2	16.7–28.6	—	—	[63]
Lion	68.4–78.7	12.1–29.8	3.4–6.5			[62]
Leopard	63.2–72.9	11.3–16.6	3.4–5.2	—	—	[62]

The ranges showed were derived from several results of the respective studies cited.

- Pregnancy rate was derived from the total pregnancies established/total transferred recipients, according to each reference.
- Live birth rate correspond to the transferred recipient that delivered healthy kittens/total transferred recipient, according to each reference.

Table 4. Efficiency of SCNT in the domestic cat and wild felid species.

The technique used for the embryo generation might also affect the gene expression pattern. A decreased *OCT4* expression was related to a reduced developmental capacity of feline embryos generated by iSCNT [68]. Moreover, it has been described that cheetah embryos generated by iSCNT have an aberrant expression of *OCT4*, *SOX2*, *NANOG* and *CDX2* compared to domestic cat embryos generated by SCNT and IVF. This was in accordance with a reduced developmental competence of the cheetah cloned embryos, which might be due to an inadequate reprogramming of the cheetah donor cell by the cat oocyte [63].

3. Cellular regenerative therapies in felids

The challenge in feline cell therapy is to offer treatment options to different diseases and to use stem cells in endangered animals is another challenge. This latter challenge is even more complicated because of the low availability of tissue to isolate stem cells in these species. However, biotechnology can bring us closer to the possibility of improving techniques for obtaining stem cells for therapy and tissue regeneration that may also contribute to the conservation of these individuals. This review tries to cover the main aspects of stem cells in domestic cats and its future application in wild cats, as well as their obtaining and characterization.

3.1. Induced pluripotent feline stem cells

The induced pluripotent stem cells (iPSCs) are generated by the induced expression of pluripotency transcription factors in cells that normally do not express. In this way, stem cells from somatic and differentiated cells can be obtained. Takahashi and Yamaka were the first to develop and implement this technique in 2006, who induced somatic cells mouse in iPSC by transfection of the cells with the transcription factors "OSKM" corresponding to transcription factor binding octamer 3/4 (Oct3/4), high protein group mobility-related gene Sry2 (Sox2), Kruppel factor 4 (Klf4) and c-Myc oncoprotein [69].

To date, there are reports of iPSC generated from endangered feline species such as snow leopard, tiger, jaguar and African serval [70, 71]. Protocols for induction to pluripotency in wild cats conclude that Nanog is a key factor in reprogramming [68]. For this reason, reprogramming cocktails included the four Yamanaka factors plus Nanog and use culture medium supplemented with LIF and SFB. When Nanog was removed from the cocktail, the reprogramming efficiency was greatly reduced and the colonies reached only up to the seventh pass (P7) [70]. On the other hand, the inclusion of Nanog made that the iPSCs colonies could be expanded *in vitro*, were positive for alkaline phosphatase and for OCT-4, NANOG and SSEA-2 proteins at the passage 14 (P14) [70]. Oct4 and Nanog endogenous were detected by RT-PCR at passage 4 and 14, indicating reprogramming and reactivation of endogenous pluripotency genes and the transgenes of Oct4, Sox2 and Nanog were silenced at the passage 14 [70]. These cells showed a good set of stem cell characteristics like teratoma formation and genetic markers for the differentiation to ectoderm, endoderm and mesoderm. A curious fact of these protocols is the use of mouse embryonic fibroblasts (MEF) as feeder and the absence of FGF2 in the culture medium. In general, regarding the culture medium, this is simpler than the used in other species and in the domestic cat embryonic stem cell (ESC) culture [72]. This could be considered advantageous to the use of the resources, but may affect the obtaining of a "naïve" phenotype in these feline iPSCs. However, there remains doubt about the pluripotency of these cells because the chimera formation could not be assessed. There is a great interest for the feline iPSCs formation because they represent a novel opportunity for the preservation of species through assisted reproduction. The reason why there are no publications of IPSCs in domestic cats is unknown. According to some researchers, there is not enough funding for research in cell therapy in cats or it is not probably an area in demand [73]. However, the results obtained in the published studies gave the impression that the techniques in this species are still complex and require a high level of standardization.

Despite all these difficulties, the therapeutic and conservation potential that can be reached with cellular reprogramming in domestic and wild felids is promising.

3.2. Mesenchymal stem cells (MSC)

Mesenchymal stem cells (MSC), also called by some authors, as mesenchymal stromal cells are unspecialized cells that have the ability to self-renew by cell division and differentiate into specialized cells [74]. The MSCs are involved in the tissue regeneration by two different mechanisms. They may contribute directly to tissue repairing by differentiation into specific cell phenotypes such as, ligaments, tendons or alternatively fibroblasts, but not necessarily exclusive. These adult stem cells have the ability to produce extracellular matrix and bioactive proteins such as growth factors, anti-apoptotic and chemotactic agents that have an important effect on the dynamic cellular, production of anabolic effects, stimulate neovascularization and additional recruitment of stem cells the site of injury. Stem cells recruited at these sites can differentiate and/or produce biologically active peptides [75]. MSCs have a fibroblast-like morphology, must be adherent to plastic, express surface molecules like CD105, CD73 and CD90 and have a negative expression of CD45, CD34, CD14 o CD11b, CD79α o CD19 and HLA-RD. In addition, the MSCs are capable to differentiate *in vitro* into osteoblasts, adipocytes and chondroblasts [74, 76]. However, characterization of MSCs in veterinary species is more difficult due to the lack of specificity or low affinity antibodies, which are generally made against human or murine antigens. These differences in the surface markers may be due to specific features of the tissue from which the MSCs were collected. In the domestic cat, some authors report differences in the expression of surface markers in the MSCs depending of the tissue from which they were isolated (**Table 5** and **Figure 2**). An additional characteristic reported in cat MSCs derived from adipose tissue is the expression of pluripotency markers as *NANOG*, *KLF4* and *OCT4* [77].

3.2.1. Tissues source of mesenchymal stem cells (MSCs) in cats

In animals and humans, different tissues represent a potential resource for obtaining MSCs. The bone marrow was the first tissue explored; one of the best sources of production and so far the most widely used in cell therapy. However, other sources such as adipose tissue, have been investigated and successfully used. In the field of veterinary medicine, depending on the target species, the tissue from which these cells are obtained is crucial and can be influenced by the weight, size and the handling of the patient. However, the method for obtaining these tissues does not always correlate with a good source of MSCs. Therefore, the investigation of other resources for obtaining MSCs in different species is important, considering that the results between species cannot be extrapolated in all cases.

3.2.1.1. Bone marrow

MSCs isolated from bone marrow (BM-MSCs) has been described in many animal species such as mice, rabbits, horses, pigs, cattle, dogs and cat [78–83]. Maciel and colleagues conducted a study of the morphology of BM-MSC in felines and they noted the predominance of two types of cells, spindled and elongated [84]. It has been reported that feline BM-MSCs have the potential to differentiate toward neuronal lineage [85]. Feline BM-MSC after the fourth passage becomes more homogeneous and express surface markers like CD44, CD105 and CD29.

	Adipose tissue	Bone marrow	Blood	Amniotic membranes	Reference
POS	CD90	CD90	CD90	CD90	[77, 86–90]
	CD44	CD44	CD44	CD44	[86–90]
	CD105	CD105		CD105	[77, 86–90]
	CD146				[77]
	CD73				[90]
NEG	CD4	CD4	CD4		[86, 90]
	CD14			CD14	[77, 87, 88]
	CD18				[89]
	CD34			CD34	[87, 88]
	CD45			CD45	[77, 87, 88]
	CD73			CD73	[77, 87]
	CD271				[77]
	MHC II	MHC II	MHC II		[77, 86, 87, 92]

Table 5. Expression markers of domestic cat mesenchymal stem cells derived from different tissues.

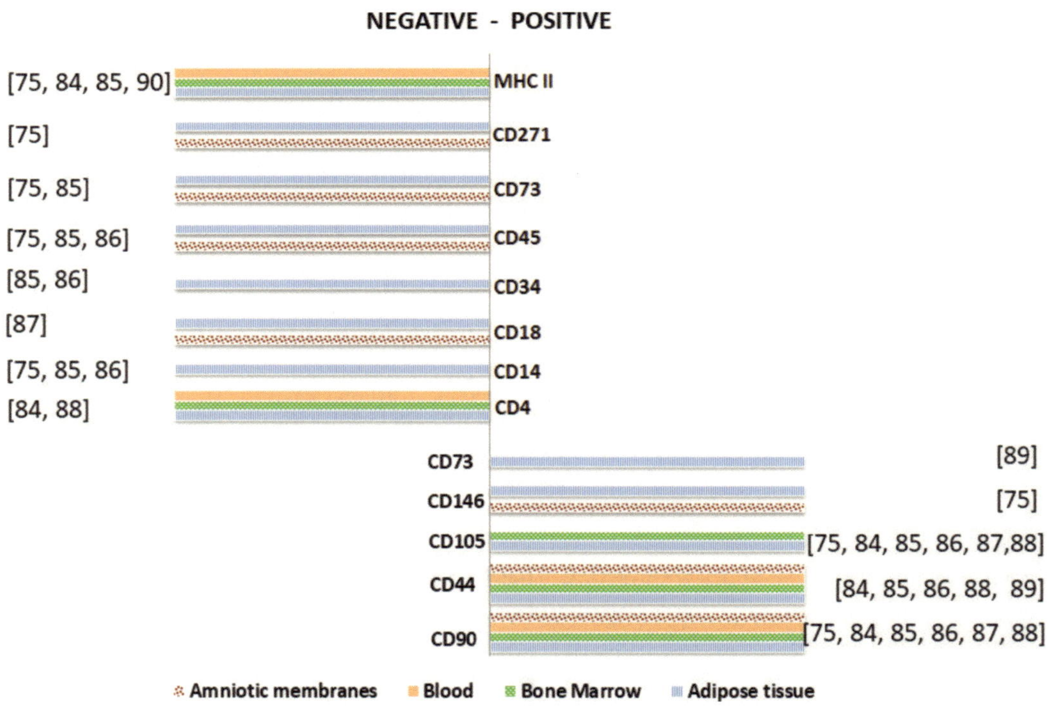

Figure 2. Representative graphic of the surface markers expression in the domestic cat mesenchymal stem cells.

3.2.1.2. *Peripheral blood*

Peripheral blood is another tissue of which has been documented obtaining of MSCs. Compared to bone marrow, it is a safer and less painful resource with less subsequent complications. Feline peripheral blood (fPBMSCs) has been isolated from venous blood of cats and expanded in culture [90]. The morphology and capacity for differentiation into mesoderm lineages of these MSCs is similar to others obtained from different feline tissues. However, these MSCs require a special culture medium containing 5% autologous plasma (AP) and 10% fetal bovine serum (FBS), showing a similar growth curve that others reports because its proliferation is limited to the seventh passage [90]. These cells express CD44 and CD90 surface markers and are negative to major histocompatibility class II and CD4.

3.2.1.3. *Fetal membranes*

The interest of MSCs derived from fetal and extra-embryonic tissues has increased because of its usefulness in the field of regenerative medicine [75, 92–94]. In felines, these MSCs are obtained from tissues discarded after birth or from procedures such as cesarean operation and ovariohysterectomy. These cells have been isolated and successfully expanded. It was reported that these cells have an adequate level of homogeneity at the passage 3, as it was found in canines [95]. Unlike other tissues, the cell population of the feline AMSC increases considerably after the passage 7, with higher values than those reported in horses and dogs, the cell number tends to increase along with number passages [95]. Additionally, the viability of these cells after cryopreservation was similar to the viability of fresh cells, which is ideal for cell banks and future applications in cell therapy [95]. AMSCs feline express CD73 and CD90 surface markers but did not express hematopoietic specific markers CD34, CD45 and CD79 [95].

3.2.1.4. *Adipose tissue*

Adipose tissue is abundant and more accessible than other tissues and is the most commonly used in cellular veterinary therapies [89]. Cellular therapies for animals using cells derived from adipose tissue (AMSC) have been used to treat osteoarthritis, injuries of ligaments and tendons in canine and equine, feline gingivostomatitis with good results and other pathologies that are being studied as chronic renal failure, asthma and chronic enteropathy in domestic cats [91, 96–98]. Feline AMSCs have been isolated and characterized from black-footed cat, an endangered feline [77]. Both cat and black-footed cat AMSCs showed potential adipogenic, osteogenic and chondrogenic differentiation, but the surface marker expression of black-footed cat AMSCs is unknown and it could be difficult to evaluate because of the specificity of the antibodies. However, these results obtained from endangered felines can be used in ARTs for their conservation such as SCNT, or potentially be used in cell therapy of individuals of the same species.

3.2.1.5. *Potential sources of MSCs*

Brain, muscle, synovial fluid, tendon, placenta and dental pulp are others potential sources for obtaining MSCs in several species [99–101]. Some of these tissues could eventually be

Specie	Doses	Administration	Pathology	Age	Treatment	Reference
domestic cat	3 doses of 2 × 10^6 cells/kg	Intravenous	Chronic renal failure	10–15 years	Allogenic	[98]
Domestic cat	2 doses of 2 × 10^6 cells/kg	Intravenous	Enteropathy	7–15 years	Allogenic	[104]
Domestic cat	1 doses of 5 × 10^6 cells/kg	Intravenous	Gingivostomatitis	1–14 years	Autologus	[91]

Table 6. Treatment by pathology, doses and administration route of MSCs in felines.

used to isolate mesenchymal stem cells in felines to perform extensive characterization procedures, which could defined if these cells have any potential to be employed in cell therapy. Considering that stem cells can be isolated for other purposes such as species conservation, spermatogonial stem cells (SSCs) can be isolated from testicular tissue. In domestic cats, these cells were successfully isolated and cultured *in vitro,* but its viability was compromised and the culture only reached 57 days [102]. Interestingly, SSCs can differentiate into sperm and be available for transplantation in testicular tissue. The first successful xenotransplant study was performed using ocelot SSCs in domestic cat testicular tissue [103]. Ocelot (*Leopardus pardalis*) spermatozoa were observed in the cat epididymis 13 weeks following transplantation suggesting that the domestic cat may be a useful species to produce gametes from another close species. This study along with others demonstrates the importance of the domestic cat as a research model for the conservation of endangered species [103].

3.3. Doses and routes of administration for MSC treatment in domestic cats

To date, mesenchymal stem cells are the only ones that have been evaluated in treatments of certain pathologies in domestic cats. The establishment of the treatment doses of MSCs in cats has been a main point to obtain the expected therapeutic effect and to reduce the adverse effects that may occur. After the adequate characterization of MSCs depending on their place of origin, the challenge is to evaluate the dose response and the route of administration. It has been reported that the intravenous route is the most chosen in the treatments with MSCs. Allogenic and autologus administration of MSCs in cats have not shown any side effects [96, 98, 104]. More studies are still required to determine alternative administration routes that may increase the treatment efficacy. Furthermore, the doses used are still a topic of discussion, but some doses are proposed for certain treatments (**Table 6**). The effects of reported treatments suggest that they can be seen since the third dose of MSCs administration [96].

4. Conclusions

Great advances have been made in the *in vitro* embryo production and cellular therapies in felids. The studies first made in the domestic cat have allowed the subsequent application of these techniques in endangered felids. Regarding the *in vitro* embryo production, several live birth reports have demonstrated the efficiency of these technique in the conservation of endangered felids. However, more studies are needed to improve their efficiency especially in the case of the SCNT. On the other hand, the results in cell therapy applied to felids set the background for future research. There is still need to identify the potential of stem cells isolated from different felid tissues. Additionally, the tools for the isolation and characterization of these cells in wild felid species must be improved considering that it is a very limited resource. However, there are unexplored alternatives for the use of stem cells, not only as therapy for different pathologies but also for obtaining gametes or other cell types from which is possible to rescue genetic material.

Author details

Daniel Veraguas, Diana Echeverry, Fidel Ovidio Castro and Lleretny Rodriguez-Alvarez*

*Address all correspondence to: llrodriguez@udec.cl

Department of Animal Science, Faculty of Veterinary Sciences, University of Concepcion, Chillán, Chile

References

[1] Gómez MC, Pope CE. Pluripotent and multipotent domestic cat stem cells: Current knowledge and future prospects. In: Brevini TAL, editor. Stem Cells in Animal Species: From Pre-Clinic to Biodiversity. Humana Press. NY, USA. Springer International Publishing; 2014. p. 119-142

[2] Pope CE, Gómez MC, Dresser BL. *In vitro* production and transfer of cat embryos in the 21st century. Theriogenology. 2006;**66**:59-71

[3] IUCN. The IUCN red list of threatened species. Version 2015-4. www.iucnredlist.org; 2016 [accessed 18.04.16].

[4] Swanson WF, Brown JL. International training programs in reproductive sciences for conservation of Latin American felids. Animal Reproduction Science. 2004;**82**:21-34

[5] Swanson WF. Application of assisted reproduction for population management in felids: The potential and reality for conservation of small cats. Theriogenology. 2006;**66**:49-58

[6] Pope CE. Embryo technology in conservation efforts for endangered felids. Theriogenology. 2000;**53**:163-174

[7] Pope CE, Gómez MC, Galiguis J, Dresser BL. Applying embryo cryopreservation technologies to the production of domestic and black-footed cats. Reproduction in Domestic Animals. 2012;**47**:125-129

[8] Pope CE, Johnson CA, McRae MA, Keller GL, Dresser BL. Development of embryos produced by intracytoplasmic sperm injection of cat oocytes. Animal Reproduction Science. 1998;**53**:221-236

[9] Gómez MC, Pope CE, Harris R, Davis A, Mikota S, Dresser BL. Births of kittens produced by intracytoplasmic sperm injection of domestic cat oocytes matured *in vitro*. Reproduction, Fertility, and Development. 2000;**12**:423-433

[10] Moro LN, Sestelo AJ, Salamone DF. Evaluation of cheetah and leopard spermatozoa developmental capability after interspecific ICSI with domestic cat oocytes. Reproduction in Domestic Animals. 2014;**49**:693-700

[11] Vajta G, Gjerris M. Science and technology of farm animal cloning: State of the art. Animal Reproduction Science. 2006;**92**:211-230

[12] Shin T, Kraemer D, Pryor J, Liu L, Rugila J, Howe L, Buck S, Murphy K, Lyons L, Westhusin MA. Cat cloned by nuclear transplantation. Nature. 2002;**415**:859

[13] Gómez MC, Jenkins JA, Giraldo A, Harris RF, King A, Dresser BL, Pope CE. Nuclear transfer of synchronized African wild cat somatic cells into enucleated domestic cat oocytes. Biology of Reproduction. 2003;**69**:1032-1041

[14] Gómez MC, Pope CE, Giraldo A, Lyons LA, Harris RF, King AL, Cole A, Godke RA, Dresser BL. Birth of African Wildcat cloned kittens born from domestic cats. Cloning and Stem Cells. 2004;**6**:247-258

[15] Farstad W. Current state in biotechnology in canine and feline reproduction. Animal Reproduction Science. 2000;**60-61**:375-387

[16] Pope CE. Aspects of *in vivo* oocyte production, blastocyst development, and embryo transfer in the cat. Theriogenology. 2014;**81**:126-137

[17] Howard J, Donoghue AM, Barone MA, Goodrowe KL, Blumer ES, Snodgrass K, Starnes D, Tucker M, Bush M, Wildt DE. Successful induction of ovarian activity and laparoscopic intrauterine artificial insemination in the cheetah (*Acinonyx jubatus*). Journal of Zoo and Wildlife Medicine. 1992;**23**:288-300

[18] Donoghue AM, Johnston LA, Armstrong DL, Simmons LG, Wildt DE. Birth of a Siberian tiger cub (*Panthera tigris altaica*) following laparoscopic intrauterine artificial insemination. Journal of Zoo and Wildlife Medicine. 1993;**24**:185-189

[19] Howard J, Byers AP, Brown JL, Barrett SJ, Evans MZ, Schwartz RJ, Wildt DE. Successful ovulation induction and laparoscopic intrauterine artificial insemination in the clouded leopard (*Neofelis nebulosa*). Zoo Biology. 1996;**15**:55-69

[20] Barone MA, Wildt DE, Byers AP, Roelke ME, Glass CM, Howard JG. Gonadotrophin dose and timing of anaesthesia for laparoscopic artificial insemination in the puma (*Felis concolor*). Journal of Reproduction and Fertility. 1994;**101**:103-108

[21] Hamner CE, Jennings LL, Sojka NJ. Cat (*Felis catus* L.) spermatozoa require capacitation. Journal of Reproduction and Fertility. 1970;**23**:477-480

[22] Goodrowe KL, Wall RJ, O'brien SJ, Schmidt PM, Wildt DE. Developmental competence of domestic cat follicular oocytes after fertilization *in vitro*. Biology of Reproduction. 1988;**39**:355-372

[23] Roth TL, Swanson WF, Wildt DE. Developmental competence of domestic cat embryos fertilized *in vivo* versus *in vitro*. Biology of Reproduction. 1994;**51**:441-451

[24] Wolfe BA, Wildt DE. Development to blastocysts of domestic cat oocytes matured and fertilized *in vitro* after prolonged cold storage. Journal of Reproduction and Fertility. 1996;**106**:135-141

[25] Sananmuang T, Tharasanit T, Nguyen C, Phutikanit N, Techakumphu M. Culture medium and embryo density influence on developmental competence and gene expression of cat embryos. Theriogenology. 2011;**75**:1708-1719

[26] Herrick JR, Bond JB, Magarey GM, Bateman HL, Krisher RL, Dunford SA, Swanson WF. Toward a feline-optimized culture medium: Impact of ions, carbohydrates, essential amino acids, vitamins, and serum on development and metabolism of *in vitro* fertilization-derived feline embryos relative to embryos grown *in vivo*. Biology of Reproduction. 2007;**76**:858-870

[27] Freistedt P, Stojkovic M, Wolf E. Efficient *in vitro* production of cat embryos in modified synthetic oviduct fluid medium: Effects of season and ovarian status. Biology of Reproduction. 2001;**65**:9-13

[28] Karja NW, Otoi T, Murakami M, Yuge M, Fahrudin M, Suzuki T. Effect of protein supplementation on development to the hatching and hatched blastocyst stages of cat IVF embryos. Reproduction, Fertility, and Development. 2002;**14**:291-296

[29] XF Y, Cho SJ, Bang JI, Lee HS, Lee YS, Kwon TH, Deb GK, Kong IK. Effect of equine chorionic gonadotropin on the efficiency of superovulation induction for *in vivo* and *in vitro* embryo production in the cat. Theriogenology. 2010;**73**:413-420

[30] Veraguas D, Gallegos PF, Velasquez AE, Castro FO, Rodriguez-Alvarez L. FSH stimulation of anestrous cats improves oocyte quality and development of parthenogenetic embryos. Theriogenology. 2017;**87**:25-35

[31] Veraguas D, Cuevas SR, Gallegos PF, Castro FO, Rodriguez-Alvarez L. 189 effect of the ovarian stimulation of anestrous cats with eCG on morphological quality and gene expression profile of cumulus-oocyte complexes. Reproduction, Fertility, and Development. 2017;**29**:203-203 (Abstract)

[32] Pope CE, Gómez MC, Dresser BL. In vitro embryo production and embryo transfer in domestic and non-domestic cats. Theriogenology. 2006;**66**:1518-1524

[33] Kanda M, Oikawa H, Nakao H, Tsutsui T. Early embryonic development *in vitro* and embryo transfer in the cat. The Journal of Veterinary Medical Science. 1995;**57**:641-646

[34] Pope CE, Keller GL, Dresser BL. *In vitro* fertilization in domestic and non-domestic cats including sequences of early nuclear events, development *in vitro*, cryopreservation and successful intra-and interspecies embryo transfer. Journal of Reproduction and Fertility. Supplement. 1993;**47**:189-201

[35] Pope CE, Gómez MC, Mikota SK, Dresser BL. Development of *in vitro* produced African wildcat (*Felis silvestris*) embryos after cryopreservation and transfer into domestic cat recipients. Biology of Reproduction. 2000;**321**(abstract):62

[36] Pope CE, Gómez MC, Cole A, Dumas C, Dresser BL. Oocyte recovery, *in vitro* fertilization and embryo transfer in the serval (*Leptailurus serval*). Reproduction, Fertility, and Development 2005;**18**:223 (abstract).

[37] Herrick JR, Mehrdadfar F, Campbell M, Levens G, Leiske K, Swanson WF. Birth of sand cat (*Felis margarita*) kittens following *in vitro* fertilization and embryo transfer. Biology of Reproduction. 2010;**83**:28 (abstract)

[38] Gañán N, González R, Sestelo A, Garde JJ, Sánchez I, Aguilar JM, Gomendio M, Roldan ERS. Male reproductive traits, semen cryopreservation, and heterologous in vitro fertilization in the bobcat (*Lynx rufus*). Theriogenology. 2009;**72**:341-352

[39] Gañán N, González R, Garde JJ, Martínez F, Vargas A, Gomendio M, Roldan ER. Assessment of semen quality, sperm cryopreservation and heterologous IVF in the critically endangered Iberian lynx (*Lynx pardinus*). Reproduction, Fertility, and Development. 2009;**21**:848-859

[40] Donoghue AM, Johnston LA, Seal US, Armstrong DL, Tilson RL, Wolf P, Petrini K, Simmons LG, Gross T, Wildt DE. *In vitro* fertilization and embryo development *in vitro* and *in vivo* in the tiger (*Panthera tigris*). Biology of Reproduction. 1990;**43**:733-744

[41] Donoghue AM, Johnston LA, Seal US, Armstrong DL, Simmons LG, Gross T, Tilson RL, Wildt DE. Ability of thawed tiger (*Panthera tigris*) spermatozoa to fertilize conspecific eggs and bind and penetrate domestic cat eggs *in vitro*. Journal of Reproduction and Fertility. 1992;**96**:555-564

[42] Crichton EG, Bedows E, Miller-Lindholm AK, Baldwin DM, Armstrong DL, Graham LH, Ford JJ, Gjorret JO, Hyttel P, Pope CE, Vajta G, Loskutoff NM. Efficacy of porcine gonadotropins for repeated stimulation of ovarian activity for oocyte retrieval and *in vitro* embryo production and cryopreservation in Siberian tigers (*Panthera tigris altaica*). Biology of Reproduction. 2003;**68**:105-113

[43] Miller AM, Roelke ME, Goodrowe KL, Howard JG, Wildt DE. Oocyte recovery, maturation and fertilization *in vitro* in the puma (*Felis concolor*). Journal of Reproduction and Fertility. 1990;**88**:249-258

[44] Donoghue AM, Howard JG, Byers AP, Goodrowe KL, Bush M, Blumer E, Lukas J, Stover J, Snodgrass K, Wildt DE. Correlation of sperm viability with gamete interaction and fertilization *in vitro* in the cheetah (*Acinonyx jubatus*). Biology of Reproduction. 1992;**46**:1047-1056

[45] Wildt DE, Bush M, Howard JG, O'Brien SJ, Meltzer D, Van Dyk A, Ebedes H, Brand DJ. Unique seminal quality in the south African cheetah and a comparative evaluation in the domestic cat. Biology of Reproduction. 1983;**29**:1019-1025

[46] Wildt DE, O'Brien SJ, Howard JG, Caro TM, Roelke ME, Brown JL, Bush M. Similarity in ejaculate-endocrine characteristics in captive versus free-ranging cheetahs of two subspecies. Biology of Reproduction. 1987;**36**:351-360

[47] Armstrong DL, Crichton EG, Dankof SM, Schwalbach LMJ, Gardner DK, Loskutoff NM. 200 ovarian stimulation, laparoscopic oocyte retrieval, ivf and blastocyst production using sequential media in the African lion (*Panthera leo*). Reproduction, Fertility, and Development. 2003;**16**:221 (abstract)

[48] Rubino P, Viganò P, Luddi A, Piomboni P. The ICSI procedure from past to future: A systematic review of the more controversial aspects. Human Reproduction Update. 2015;**22**:194-227

[49] Wildt DE, Phillips LG, Simmons LG, Chakraborty PK, Brown JL, Howard JG, Teare A, Bush MA. Comparative analysis of ejaculate and hormonal characteristics of the captive male cheetah, tiger, leopard, and puma. Biology of Reproduction. 1988;**38**:245-255

[50] Bogliolo L, Leoni G, Ledda S, Naitana S, Zedda M, Carluccio A, Pau S. Intracytoplasmic sperm injection of *in vitro* matured oocytes of domestic cats with frozen-thawed epididymal spermatozoa. Theriogenology. 2001;**56**:955-967

[51] Damiani P, Gomez M, Cole A, Pope E, Aguilar R, Hammond B, Nel L, Cortez C, Vacarro J, Sarrat E, Markey E, Dresser B. 204 the production of intracytoplasmic sperm injection lion (*Panthera leo*) embyros using spermatozoa collected by percutaneous epididymal sperm aspiration from vasectomized males. Reproduction, Fertility, and Development. 2003;**16**:223-224

[52] Fernandez-Gonzalez L, Hribal R, Stagegaard J, Zahmel J, Jewgenow K. Production of lion (*Panthera leo*) blastocysts after *in vitro* maturation of oocytes and intracytoplasmic sperm injection. Theriogenology. 2015;**83**:995-999

[53] Niemann H. Epigenetic reprogramming in mammalian species after SCNT-based cloning. Theriogenology. 2016;**86**:80-90

[54] Yin XJ, Lee HS, Lee YH, Seo YI, Jeon SJ, Choi EG, Cho SJ, Cho SG, Min W, Kang SK, Hwang WS, Kong IK. Cats cloned from fetal and adult somatic cells by nuclear transfer. Reproduction. 2005;**129**:245-249

[55] Yin XJ, Lee HS, XF Y, Choi E, Koo BC, Kwon MS, Lee YS, Cho SJ, Jin GZ, Kim LH, Shin HD, Kim T, Kim NH, Kong IK. Generation of cloned transgenic cats expressing red fluorescence protein. Biology of Reproduction. 2008;**78**:425-431

[56] Gómez MC, Pope CE, Kutner RH, Ricks DM, Lyons LA, Ruhe MT, Dumas C, Lyons J, Dresser BL, Reiser J. Generation of domestic transgenic cloned kittens using lentivirus vectors. Cloning and Stem Cells. 2009;**11**:167-176

[57] Gómez MC, Pope CE, Kutner RH, Ricks DM, Lyons LA, Ruhe M, Dumas C, Lyons J, López M, Dresser BL, Reiser J. Nuclear transfer of sand cat cells into enucleated domestic cat oocytes is affected by cryopreservation of donor cells. Cloning and Stem Cells. 2008;**10**:469-484

[58] Yin X, Lee Y, Lee H, Kim N, Kim L, Shin H, Kong I. *In vitro* production and initiation of pregnancies in inter-genus nuclear transfer embryos derived from leopard cat (*Prionailurus bengalensis*) nuclei fused with domestic cat (*Felis silverstris catus*) enucleated oocytes. Theriogenology. 2006;**66**:275-282

[59] Gómez MC, Pope CE, Ricks DM, Lyons J, Dumas C, Dresser BL. Cloning endangered felids using heterospecific donor oocytes and interspecies embryo transfer. Reproduction, Fertility, and Development. 2009;**21**:76-82

[60] Thongphakdee A, Siriaroonrat B, Manee-in S, Klincumhom N, Kamolnorranath S, Chatdarong K, Techakumphu M. Intergeneric somatic cell nucleus transfer in marbled cat and flat-headed cat. Theriogenology. 2010;**73**:120-128

[61] Hashem M, Bhandari DP, Kang SK, Lee BC. Cell cycle analysis and interspecies nuclear transfer of *in vitro* cultured skin fibroblasts of the Siberian tiger (*Panthera tigris Altaica*). Molecular Reproduction and Development. 2007;**74**:403-411

[62] Yelisetti UM, Komjeti S, Katari VC, Sisinthy S, Brahmasani SR. Interspecies nuclear transfer using fibroblasts from leopard, tiger, and lion ear piece collected postmortem as donor cells and rabbit oocytes as recipients. In Vitro Cellular & Developmental Biology. Animal. 2016;**52**:632-645

[63] Moro LN, Hiriart MI, Buemo C, Jarazo J, Sestelo A, Veraguas D, Rodriguez-Alvarez L, Salamone DF. Cheetah interspecific SCNT followed by embryo aggregation improves *in vitro* development but not pluripotent gene expression. Reproduction. 2015;**150**:1-10

[64] Moro LN, Jarazo J, Buemo C, Hiriart MI, Sestelo A, Salamone DF. Tiger, Bengal and domestic cat embryos produced by Homospecific and interspecific Zona-free nuclear transfer. Reproduction in Domestic Animals. 2015;**50**:849-857

[65] Hamatani T, Carter MG, Sharov AA, Ko MSH. Dynamics of global gene expression changes during mouse preimplantation development. Developmental Cell. 2004;**6**:117-131

[66] Wrenzycki C, Herrmann D, Lucas-Hahn A, Lemme E, Korsawe K, Niemann H. Gene expression patterns in in vitro-produced and somatic nuclear transfer-derived preimplantation bovine embryos: Relationship to the large offspring syndrome? Animal Reproduction Science. 2004;**82-83**:593-603

[67] Filliers M, Goossens K, Van Soom A, Merlo B, Pope CE, De Rooster H, Smits K, Vandaele L, Peelman LJ. Gene expression profiling of pluripotency and differentiation-related markers in cat oocytes and preimplantation embryos. Reproduction, Fertility, and Development. 2012;**24**:691-703

[68] Imsoonthornruksa S, Lorthongpanich C, Sangmalee A, Srirattana K, Laowtammathron C, Tunwattana W, Somsa W, Ketudat-Cairns M, Parnpai R. Abnormalities in the transcription of reprogramming genes related to global epigenetic events of cloned endangered felid embryos. Reproduction, Fertility, and Development. 2010;**22**:613-624

[69] Takahashi K, Yamanaka S. Induction of pluripotent stem cells from mouse embryonic and adult fibroblast cultures by defined factors. Cell. 2006;**126**:663-676

[70] Verma R, Liu J, Holland MK, Temple-Smith P, Williamson M, Verma PJ. Nanog is an essential factor for induction of pluripotency in somatic cells from endangered felids. BioResearch Open Access. 2013;**2**:72-76

[71] Verma R, Holland MK, Temple-Smith P, Verma PJ. Inducing pluripotency in somatic cells from the snow leopard (*Panthera uncia*), an endangered felid. Theriogenology. 2012;**77**:220-228

[72] Gómez M, Serrano M, CE P, Jenkins J, Biancardi MN, López M, Dumas C, Galiguis J, Dresser BL. Derivation of cat embryonic stem-like cells from *in vitro*-produced blastocysts on homologous and heterologous feeder cells. Theriogenology. 2010;**74**:679-688

[73] Paterson YZ, Kafarnik C, Guest DJ. Characterization of companion animal pluripotent stem cells. Cytometry. Part A. 2017;**7**:1-12

[74] Caplan AI. Mesenchymal stem cells. Journal of Orthopaedic Research. 1991;**9**:641-650

[75] Dahlgren LA. Fat-derived mesenchymal stem cells for equine tendon repair. In: World Conference on Regenerative Medicine. Regen Med Suppl 2009;**4** (Suppl. 2).

[76] Denham M, Conley B, Olsson F, Cole T, Mollard R. Stem cells: An overview. Current Protocols in Cell Biology. 2005;**23**:23

[77] Gómez MC, Qin Q, Biancardi MN, Galiguis J, Dumas C, MacLean R, Wang G, Pope CE. Characterization and multilineage differentiation of domestic and black-footed cat mesenchymal stromal/stem cells from abdominal and subcutaneous adipose tissue. Cellular Reprogramming. 2015;**17**:376-392

[78] Bosnakovski D, Mizuno M, Kim G, Takagi S, Okumura M, Fujinaga T. Isolation and multilineage differentiation of bovine bone marrow mesenchymal stem cells. Cell and Tissue Research. 2005;**319**:243-253

[79] Im GI, Kim DY, Shin JH, Hyun CW, Cho WH. Repair of cartilage defect in the rabbit with cultured mesenchymal stem cells from bone marrow. Journal of Bone and Joint Surgery. British Volume (London). 2001;**83**:289-294

[80] Martin DR, Cox NR, Hathcock TL, Niemeyer GP, Baker HJ. Isolation and characterization of multipotential mesenchymal stem cells from feline bone marrow. Experimental Hematology. 2002;**30**:879-886

[81] Ringe J, Kaps C, Schmitt B, Buscher K, Bartel J, Smolian H, Schultz O, Burmester GR, Häupl T, Sittinger M. Porcine mesenchymal stem cells. Induction of distinct mesenchymal cell lineages. Cell and Tissue Research. 2002;**307**:321-327

[82] Tropel P, Noël D, Platet N, Legrand P, Benabid AL, Berger F. Isolation and characterisation of mesenchymal stem cells from adult mouse bone marrow. Experimental Cell Research. 2004;**295**:395-406

[83] Vidal MA, Kilroy GE, Lopez MJ, Johnson JR, Moore RMGJ. Characterization of equine adipose tissue-derived stromal cells: Adipogenic and osteogenic capacity and comparison with bone marrow-derived mesenchymal stromal cells. Veterinary Surgery. 2007;**36**:613-622

[84] Maciel BB, Rebelatto CLK, Brofman PRS, Brito HFV, Patricio LFL, Cruz MA, Locatelli-Dittrich R. Morphology and morphometry of feline bone marrow-derived mesenchymal stem cells in culture. Pesquisa Veterinária Brasileira. 2014;**34**:1127-1134

[85] Munoz JL, Greco SJ, Patel SA, Sherman LS, Bhatt S, Bhatt RS, Shrensel JA, Guan YZ, Xie G, Ye JH, Rameshwar P, Siegel A. Feline bone marrow-derived mesenchymal stromal cells (MSCs) show similar phenotype and functions with regards to neuronal differentiation as human MSCs. Differentiation. 2012;**84**:214-222

[86] Webb TL, Quimby JM, Dow SW. In vitro comparison of feline bone marrow-derived and adipose tissue-derived mesenchymal stem cells. Journal of Feline Medicine and Surgery. 2012;**14**:165-168

[87] Iacono E, Cunto M, Zambelli DRF, Tazzari PL, Merlo B. Could fetal fluid and membranes be an alternative source for mesenchymal stem cells (MSCs) in the feline species? A preliminary study. Veterinary Research Communications. 2012;**36**:107-118

[88] Kono S, Kazama T, Kano K, Harada K, Uechi M, Matsumoto T. Phenotypic and functional properties of feline dedifferentiated fat cells and adipose-derived stem cells. Veterinary Journal. 2014;**199**:88-96

[89] Arzi B, Kol A, Murphy B, Walker NJ, Wood JA, Clark K, Verstraete FJ, Borjesson DL. Feline foamy virus adversely affects feline mesenchymal stem cell culture and expansion: Implications for animal model development. Stem Cells and Development. 2014; **24**:814-823

[90] Sato K, Yamawaki-Ogata A, Kanemoto I, Usui A, Narita Y. Isolation and characterisation of peripheral blood-derived feline mesenchymal stem cells. Veterinary Journal. 2016;**216**:183-188

[91] Arzi B, Mills-Ko E, Verstraete FJ, Kol A, Walker NJ, Badgley MR, Fazel N, Murphy WJ, Vapniarsky N, Borjesson DL. Therapeutic efficacy of fresh, autologous mesenchymal stem cells for severe refractory gingivostomatitis in cats. Stem Cells Translational Medicine. 2016;**5**:1-12

[92] Mumaw JL, Schmiedt CW, Breidling S, Sigmund A, Norton NA, Thoreson M, Peroni JF, Hurley DJ. Feline mesenchymal stem cells and supernatant inhibit reactive oxygen species production in cultured feline neutrophils. Research in Veterinary Science. 2015; **103**:60-69

[93] Csaki C, Matis U, Mobasheri A, Ye H, Shakibaei M. Chondrogenesis, osteogenesis and adipogenesis of canine mesenchymal stem cells: A biochemical, morphological and ultrastructural study. Histochemistry and Cell Biology. 2007;**128**:507-520

[94] Hoynowski S, Fry M, Gardner B, Leming M, Tucker J, Black L, Sand T, Mitchell KE. Characterization and differentiation of equine umbilical cordderived matrix cells. Biochemical and Biophysical Research Communications. 2007;**362**:347-353

[95] Vidane AS, Souza AF, Sampaio RV, Bressan FF, Pieri NC, Martins DS, Meirelles FV, Miglino MA, Ambrósio CE. Cat amniotic membrane multipotent cells are nontumorigenic and are safe for use in cell transplantation. Stem Cells Cloning. 2014;**7**:71-78

[96] Trzil JE, Masseau I, Webb TL, Chang C-H, Dodam JR, Liu H, Quimby JM, Dow SW, Reinero CR. Intravenous adipose-derived mesenchymal stem cell therapy for the treatment of feline asthma: A pilot study. Journal of Feline Medicine and Surgery. 2015;**18**:981-990

[97] Webb TL, Webb CB. Stem cell therapy in cats with chronic enteropathy: A proof-of-concept study. Journal of Feline Medicine and Surgery. 2015;**17**:901-908

[98] Quimby JM, Webb TL, Randall E, Marolf A, Valdes-Martinez A, Dow SW. Assessment of intravenous adipose-derived allogeneic mesenchymal stem cells for the treatment of feline chronic kidney disease: A randomized, placebo-controlled clinical trial in eight cats. Journal of Feline Medicine and Surgery. 2015;**18**:165-171

[99] Durgam SS, Stewart AA, Caporali EH, Karlin WM, Stewart MC. Effect of tendonderived progenitor cells on a collagenase-induced model of tendinitis in horses. Regenerative Medicine Supplement. 2009;**4**(6)

[100] Jiang Y, Jahagirdar BN, Reinhardt RL, Schwartz RE, Keene CD, Ortiz-Gonzalez XR, Reyes M, Lenvik T, Lund T, Blackstad M, Du J, Aldrich S, Lisberg A, Low WC, Largaespada DA, Verfaillie CM. Pluripotency of mesenchymal stem cells derived from adult marrow. Nature. 2002;**418**:41-49

[101] Stewart A, Chen YJ, Caporali E, Stewart A. Isolation and chondrogenic differentiation of cells isolated from equine synovial fluid. In: World Conference on Regenerative. Med Regen Med Suppl 2009;**4**(6).

[102] Tiptanavattana N, Thongkittidilok C, Techakumphu M. Characterization and in vitro culture of putative spermatogonial stem cells derived from feline testicular tissue. The Journal of Reproduction and Development. 2013;**59**:189-195

[103] Silva RC, Costa GM, Lacerda SM, Batlouni SR, Soares JM, Avelar GF, Böttger KB, Silva SF Jr, Nogueira MS, Andrade LM, França LR. Cell transplantation in felids: A potential approach to preserving endangered species. Journal of Andrology. 2012;**33**:264-276

[104] Quimby JM, Webb TL, Habenicht LM, Dow SW. Safety and efficacy of intravenous infusion of allogeneic cryopreserved mesenchymal stem cells for treatment of chronic kidney disease in cats: Results of three sequential pilot studies. Stem Cell Research & Therapy. 2013;**4**:48

[105] Gómez MC, Pope E, Harris R, Mikota S, Dresser BL. Development of in vitro matured, in vitro fertilized domestic cat embryos following cryopreservation, culture and transfer. Theriogenology. 2003;**60**:239-251

Chapter 5

Genetic Characterization of Jaguars (*Panthera onca*) in Captivity in Zoological Parks of Colombia

Santiago Jimenez Gonzalez, Manuel Ruiz García, Jesús Maldonado, Ligia Jimenez, Alberto Sánchez, Tammy Wilbert, Susan L. Castro, Jimmy H. Vargas, Sonia R. Quintanilla, Haydy Monsalve and Claudia Jimenez Escobar

Additional information is available at the end of the chapter

http://dx.doi.org/10.5772/intechopen.69779

Abstract

The construction of the pedigree of captive jaguars (*Panthera onca*) in zoological parks of Colombia was done using the analysis of the Regional Studbook for Jaguars and DNA analysis of 9 microsatellites of 20 Jaguars (n=20). The assignments for paternities and maternities were done with for the program CERVUS and the relationship between animals were established with the KINSHIP program. The analysis of the Studbook was done with SPARKS and PM2000 software generating the following values: genetic diversity for the population (GD=0.7832), potential genetic diversity (GD=0.9113), genic value (GV=0.7846), mean coefficient of inbreeding (F=0.0179), and the Mean KINSHIP (MK) for each individual. The averages of the observed and expected heterozygosity were 0.687 and 0.684 respectively. Nevertheless, a wild jaguar sample of 156 individuals obtained in Colombia substantially showed a higher degree of gene diversity (H = 0.87) than the Colombian captive jaguar population. Thus, the captive jaguar population retained 78 % of the gene diversity of the Colombian wild jaguar population. With this study the pedigree of the captive population of jaguars was built in order to develop an ex situ conservation plan for the species in the Colombian zoological parks.

Keywords: jaguar, *Panthera onca*, genetics, microsatellites, DNA, Studbook

INTECH
open science | open minds

1. Introduction

During the last decades, genetic analyses have been conducted using microsatellites, on different species of wild felids, in order to solve questions regarding the origin and evolution of members of the *Felidae* family. DNA microsatellites are composed of tandem repetitive units of two to six base pairs in length [1]. Microsatellites are randomly distributed, highly polymorphic, and frequently found inside eukaryotic genomes. One of the first reports about the use of microsatellites was done in order to determine the time when the first great depletion of cheetahs (*Acinonyx jubatus*) was proven, as evidence was found, of a "bottle neck" effect in the population, approximately 10,000 years ago [2]. Subsequently, they were employed by the same group of investigators, to determine the level of genetic diversity in four species of felids: domestic cats (*Felis catus*), cheetahs, pumas (*Puma concolor*), and African lions (*Panthera leo*) [3]. In 2001 a study was conducted on pumas that enabled to confirm the presence of homoplasy in 10 microsatellites in pumas [4]. Microsatellites have also been employed to detect differences among subspecies of clouded leopards (*Neofelis nebulosa nebulosa*) found in Thailand and the clouded leopard (*N. nebulosa diardi*) found in the Island of Borneo [5]. In this way, microsatellites have also become a fundamental tool to differentiate subspecies of captive tigers (*Panthera tigris*), at world level, that could be employed as important genetic reservoirs [6]. Also microsatellites have been applied to different Neotropical cat species, several studies have been carried out with the ocelot (*Leopardus pardalis*). These studies with microsatellites revealed the non-existence of agreement between the putative morphological subspecies of ocelots and the molecular results obtained [7, 8]. Contrarily, microsatellites have been usefully employed in the expansion determination of significant molecular subspecies or evolutionary units in the pampas cat, *Leopardus pajeros*, and in the Andean cat, *L. jacobita* [9–11]. As for jaguars (*Panthera onca*), the first analysis with microsatellites employed 29 loci analyzed, finding molecular evidence of a recent population expansion and suggesting differences due to the geographic barriers like the *Tapon del Darien* and the Amazon river [12]. Nevertheless, more recent studies with jaguars and microsatellites have shown the non-existence of significant geographical barriers to the dispersion of the jaguars [13, 14]. Likewise, the variance in the size of the alleles of the microsatellites has permitted to develop genomic estimates for phenomena that occurred in ancestral times, such as the founding effects presented in the North American pumas, Asiatic lions (*P. leo persicus*), and cheetah [15], as well as some studies have shown the existence of mutation constrictions or some selective pressures on the microsatellite evolution in felids [16]. The first reports about the phyllogeography and the natural history of jaguars were conducted with the analysis of microsatellites and of mitochondrial DNA by Eizirik [12]. The employment of these is suggested to characterize specimens of different origins, thus determining possible crossbreeding of specimens in captivity, which represent the same genetic legacy. In a subsequent study, great genetic diversity was reported and a small genetic heterogeneity between the subspecies was found in Colombia, composed mainly by Central American jaguars (*P. onca centralis*) and Amazonic jaguars (*P. onca onca*) [13]. Lately, other works determined the non-existence of bottlenecks, different temporal splits in the diversification of mitochondrial lineages, and some estimation of effective numbers in the jaguar populations of Colombia and other Latin American regions [7, 14].

In Colombia, the genetic structure of jaguars held captive in zoological parks is unknown, as is their pedigree, their importance as possible genetic reservoir, and their feasible participation in a conservation plan *ex situ*, which might consider the viability and the vitality of the population at a genetic level as well as reproductive. Therefore, the main aims of the current work are as follows: (1) Conducting a retrospective genetic analysis from the records of the regional jaguar (*P. onca*) Studbook for Colombia and knowing the current genetic characteristics of the population in captivity, future sustainable projections can be established at a reproductive level, for the captive population. (2) The construction of a pedigree of the captive Colombian jaguar population through the identification of individuals, using microsatellites, could help to determine the genetic structure of this captive population. It could be useful to identify specimens which have a high genetic and reproductive potential in a conservation plan that may contribute to the conservation *ex situ*, in the zoological parks of Colombia, as a genetic reservoir, and in the same way that may contribute to the conservation *in situ* of the species in their natural habitat. (3) We compared the gene diversity levels found in the Colombian jaguar captive population with the gene diversity found in wildlife jaguars sampled in Colombia.

2. Materials and methods

2.1. Genetic analysis with microsatellites

2.1.1. Samples

Seeking to construct the pedigree of the population, blood samples were obtained from captive jaguars (n = 15) under general anesthesia, in three Colombian zoological parks: Fundación Zoológico Santacruz (n = 8) located in the San Antonio del Tequendama municipality in the department of Cundinamarca with coordinates Latitude 6°13′23.93″N Longitude 75°34′48.81″W, Zoológico Matecaña (n = 6) located in the city of Pereira in the department of Risaralda with coordinates Latitude 4°49′0.30″N Longitude 75°44′15.04″W, Parque Recreativo y Zoológico Piscilago (n = 1) located in the Nilo Municipality in the department of Cundinamarca coordinates Latitude 11° 0′39.84″N Longitude 74°47′53.69″W; hair follicles were also obtained from the jaguars in the Fundación Zoológico Santafé (n = 5) located in the city of Medellin, in the department of Antioquia with coordinates Latitude 6°13′23.93″N Longitude 75°34′48.81″W. The information to complete the construction of the pedigree, regarding the animals present in the Piscilago Zoological Park (n = 2), Barranquilla Zoological Park (n = 3), Bioparque los Ocarros (n = 2), Fundación Zoológico de Cali (n = 2), and Fundación Zoológico Parque Jaime Duque (n = 3), was obtained from the Regional Jaguar Studbook for Colombia.

The samples of the 156 jaguars sampled in the wild in Colombia for comparative purposes were from the following localities as follows: representing the putative subspecies *centralis*, the samples were from the Departments of Atlantico, Bolivar (Colorado), Magdalena (Sierra Nevada de Santa Marta, Antioquia (Apartadó, Santa Fé de Antioquia), Norte de Santander, Chocó (St. Maria de Condoto, Utría), Risaralda and Cauca, and representing the putative subspecies *onca*, the samples were from the Departments of Arauca (Lipa River), Meta (La Macarena), Caquetá (Caguán River), Vichada (Tuparro National Park), Guainía (Inirida River, Isana River), Vaupés (Miraflores, Itilla River, Apaporis River, Yuruparí), Guaviare (San José del Guaviare, Itilla River, Guayabero River), Putumayo and Amazonas (from Leticia to San Juan de Atacuarí).

2.1.2. *Anesthesia*

Following a 24-h fast, an anesthesia protocol was conducted, using a combination of xylazine–ketamine. Xylazine, an agonist alfa 2 adrenergic sedative, was administered with an initial injection. After 5 min of its application, and once signs of sedation were observed, a second injection with a fixed dissociative anesthetic, derived from cyclohexylamines (ketamine), was administered. The administration of these medications was conducted using devices of drug injections at a distance (DIDD). The site of the impact of the dart was localized in the muscular mass of the upper and lower limbs; the former being preferred, seeking that the dart or the injection be applied in the most caudal muscular mass and avoiding a possible impact on the femoral bone or the sciatic nerve. The doses of the medications administered were calculated, bearing in mind the reports on weight recorded on the medical history and the doses reported for jaguars [17, 18].

2.1.3. *Sample collection*

Blood samples were obtained by puncturing the femoral vein or the saphenous vein, having previously disinfected the area with alcohol. The samples were collected in vials with ethylenediaminetetraacetic acid (EDTA) (10 ml) and were maintained at 4°C, until their arrival to the laboratory.

2.1.4. *Microsatellites employed*

The microsatellite primers selected were developed for domestic cats [3] (**Table 1**). The criteria for the selection for the microsatellites were based on the fact that all microsatellites should be in different chromosomes and that their employment be reported on jaguars [12, 13]. The microsatellites selected were marked by fluorescence with fluorochromes that do not overlap among themselves, depending on the microsatellite selected. Only the first of the forward sequence was fluoro-marked for the wild jaguars analyzed for comparative purposes, 12 microsatellites were employed, being them, *Fca08, Fca24, Fca43, Fca45, Fca96, Fca126, Fca136, Fca176, Fca225, Fca294, Fca391,* and *Fca506*.

2.1.5. *Extraction and quantification of DNA*

The extraction of DNA was conducted from leucocytes isolated from samples of blood, through the salting-out technique. The concentration of DNA was quantified through spectrophotometry at 260 nm. As for the temperature for each microsatellite, it was calculated through the thermodynamic formula, obtaining that the average annealing temperature needed to conduct the coupling of bases was of 55.5°C. As for the hair samples, the extraction kit DNeasy Blood and Tissue of QIAGEN was employed, following the steps given.

Name	Forward	Reverse	Fluorochrome	Temperature °C	Molecular weight NCBI
FCA075	ATGCTAATCAGTGGCATTTGG	GAACAAAAATTCCAGACGTGC	PET	57.0	104-146
FCA043	AGACGGGATTGCATGAAAAG	GAGCCACCCTAGCACATATACC	6-FAM	57.0	106-128
FCA441	ATCGGTAGGTAGGTAGATATAG	GCTTGCTTCAAAATTTTCAC	VIC	58.0	113–137
FCA008	ACTGTAAATTTCTGAGCTGGCC	TGACAGACTGTTCTGGGTATGG	NED	57.0	114-148
FCA224	CTGGGTGCTGACAGCATAGA	TGCCAGAGTTGTATGAAAGGG	6-FAM	57.0	148-180
FCA096	CACGCCAAACTCTATGCTGA	CAATGTGCCGTCCAAGAAC	NED	58.0	180–220
FCA736	TCAATGTCTTGACAACGCATAA	AGGATTGCATGACCAGGAAC	6-FAM	53.0	196-280
FCA220	CGATGGAAATTGTATCCATGG	GAATGAAGGCAGTCACAAACTG	VIC	57.0	210-224
FCA391	GCCTTCTAACTTCCTTGCAGA	TTTAGGTAGCCCATTTTCATCA	NED	57.0	222-238

Table 1. Microsatellites employed in the captive population of jaguars.

2.1.6. Polymerase chain reaction

The reaction for all the markers were carried to a total volume of 20 μl; these consisted of 1.5 mM of $MgCl_2$; 1.25 mM for each one of the dNTPs (dATP dGTP dCTP y dTTP) plus 0.4 units of DNA polymerase *Thermus thermophilus* (Tth) and 10 pmol of each first, 0.5 mg/ml of BSA just as 4 μl of genomic DNA (10 ng). The samples were amplified under the following conditions: initial denaturizing at 95°C for 5 min; 35 cycles of denaturizing at 95°C for 1 min; banding temperature corresponding to each marker for 2 min; extension at 72°C for 2 min and later 1 cycle of final extension at 72°C for 5 min employing a polymerase chain reaction (PCR) Multigene Gradient (Labnet, International) Thermal Cycler.

2.1.7. Agarose

After having conducted the PCR procedures, the products of amplification were evaluated in agarose gel at 1.5% stained with SYRB-GREEN (Invitrogen, USA). Once the amplification of the products was verified by direct visualization, they were taken to electrophoresis in capillary.

2.1.8. Electrophoresis in capillary

Electrophoresis in capillary was conducted in ABI PRISM® 310 Genetic Analyzer. The results of the electrophoresis in capillary and the naming of the alleles are determined by direct visualization of the chromatograms in the Softgenetics Gene Marker Version 1.97 program. All samples were amplified and Genotypified at least twice to minimize problems of no assignation.

2.2. Statistical analysis

2.2.1. Genetic analysis from records

Records from captive jaguars were analyzed with the information compiled in the Regional Studbook for Jaguars (*P. onca*) for Colombia, through the Single Population Analysis and

Record Keeping System (SPARKS 1.5) program, developed by the International Species Information System (ISIS). Later, seeking to conduct the genetic analysis from records, information was exported and analyzed in the program Population Management 2000 (PM2000) [19].

Statistics of individual jaguars, founders of the population, and of their descendants was calculated. The allelic **retention**, which is defined as the probability that a gene originated from a founding animal could be found in the existing current population, was calculated. The current **value of gene diversity (GD)** of the population that corresponds to the heterozygosity expected, and which is defined as the variance of allelic frequencies in a determined locus and is equal to the heterozygosity expected for a population in balance. Hardy-Weinberg with coupling of gametes at random was obtained [20–22]. The gene diversity level found in the Colombian captive jaguar population at the zoos of this country was compared with the gene diversity level obtained for the set of wildlife Colombian jaguars analyzed by Ruiz-García et al. [14]. For this, the differences among estimates were statistically analyzed with a Student *t*-test. The heterozygosity data were arcsine transformed prior to analysis, as proposed by Archie [23].

Records from captive jaguars were analyzed with the information compiled in the Regional Studbook for Jaguars (*P. onca*) for Colombia, through the Single Population Analysis and Record Keeping System (SPARKS 1.5) program, developed by the International Species Information System (ISIS). Later, seeking to conduct the genetic analysis from records, information was exported and analyzed in the program Population Management 2000 (PM2000) [19].

The **average inbreeding coefficient** ***F*** for the population was calculated over time, being this the descent of the heterozygosity observed, relative to the heterozygosity expected of the founding population [24]. The **genetic value (GV)** corresponding to the heterozygosis expected in the following generations, providing all animals reproduce and have a progeny equal to the values expected, based on the **fertility rate (Mx), mortality rate (Px)**, and **reproductive value (Vx)**, was also obtained [19].

The value of the **equivalent of the current and potential founding genome** was found; this is defined as the number of founders equally represented in the population without the random loss of alleles in the offspring awaited to produce the same genetic diversity in the population under study. Likewise, the value of the **equivalent of the current surviving founding and potential genome** was obtained [25].

Individual statistics of captive jaguars with a known pedigree were calculated, and the average kinship **(MK)** is defined as the average of coefficients of relationships that exist between an individual (including itself) and the live population of animals born in captivity. The average of kinship is equal to the proportional loss of genetic diversity of the offspring born in captivity relative to the founders. It was also calculated: the **value of kinship (KV)**, which forecasts the loss of diversity of the awaited gene of the following generation, if all animals were reproduced at random and all would produce the number of offspring awaited for each kind of age; **total genome singularity** (GU–all genome uniqueness), which is the possibility of taking an allele at random from an individual not present and that this allele be identical by means of descendants, in another live individual in the population. **Descendants genome singularity** (genome singularity relative to the population of descendants of non-founding animals); **probability of loss of genome singularity, equivalent of first-degree relatives (FOKE**, First-Order Kin Equivalent), which is the number of first-degree relatives (siblings, offspring) that should contain the number of copies of the alleles of the individuals (identical by progeny) present in the population born in captivity. For example: a son or a brother contributes 1 to FOKE; each grandson contributes ½ to FOKE; each cousin contributes ¼ to FOKE. FOKE being = 4*N*MK. The size of the effective population or **effective population number (Ne)** based on the history of the population was also calculated; the **current population number (*N*)** shows the number of live animals in the population, including males as well as females [19].

2.2.2. Genetic analysis from DNA

For the analysis of the information obtained, the *Genepop on the web* was employed [26, 27]. For each microsatellite the allelic frequencies, heterozygosity expected and observed, estimate of balance Hardy-Weinberg, linkage imbalance, determination of inbreeding through F_{is}. Calculations were performed. For assignation of paternity and maternity, and for the construction of the pedigree in the population, the programs CERVUS 3.0.3 version and KINSHIP 1.2. were employed. For the final construction of the pedigree, the following variables were taken into account: information obtained from the Studbook (gender, date of birth, previous knowledge of fathers and mothers), values of natural logarithm of relations of verisimilitude (LOD) assigned by CERVUS, with flexible intervals of assurance of 80% and strict to 95%. Likewise, the probabilities of relationship and family kinship assigned by KINSHIP for H_1 Rp = 0.5 and Rm = 0.5 versus H_0 Rp = 0 and Rm = 0 (α = 0.05, 0.01 and 0.001) where Kinship is used to prove the hypothesis of relationship of pedigree among individuals, employing data of analysis of genetic markers, after specifying the hypothesis of relationships using two variables Rp and Rm and specifying the different levels of significance Rp and Rm. The variables define the probabilities that a couple share an allele by direct line of descent coming from their father or their mother, respectively. For example, if the hypothetical relationship of two complete siblings, both values would be of 0.5. After giving the hypothesis, Kinship uses the values *r*, the allelic frequencies, and the genotypes of two individuals under consideration, to calculate the verisimilitude that such combination of genotypes could have been produced as it had been specified [28].

3. Results

3.1. Genetic analysis from the regional Studbook of jaguars for Colombia

3.1.1. Demographic and individual statistics

The demographic statistics that determine the genetic characteristics of the population of captive jaguars in Colombia were calculated. By means of the genetic analysis from records, conducted with the PM2000 program, we found that the origin of 28.9% of the population starting in the year 1968 is known, and that there have been five founding animals (T1, T2, T5, T6, T27) for the population; likewise, three possible potential founders were the individuals T32 and T33 present at the Barranquilla zoo, and the T20 in the Piscilago zoological park. These founding animals, coming from wildlife, were either captured or seized and would represent the closest genetics to that that could be found among jaguars in the wild. In this way, individual statistics were obtained, of the jaguars founders of the population, where the number is presented on the Studbook, as well as the gender, age, representation (proportion of genes within the direct line living descendants of that founder of the population), contribution (number of copies of the founder's genome that are present in the living descendants; each new generation of offspring contributes 0.5, each generation of grandchildren 0.25, etc.), allelic retention (the probability that a gene present in a founding individual exist in the living descendant animals), potential retention, and descendants. The jaguars T1 and T2 are the oldest for which there is any kind of information, they were part of the Santafé Zoo and their remains are kept in the museum of natural history of the Universidad de Antioquia; their genetic representation in the population is of 8.9% and have contributed with two descendants. Jaguars T5 and T6 are animals that remained in the Santacruz Zoological Foundation, until their death. Their genetic representation in the population is of 28 and 46% and has contributed with four and six descendants respectively. Jaguar T2, situated at the Jaime Duque Park, has a genetic representation of 7% in the population, and has contributed with a descendant.

It was also found that five founding animals have produced descendants and that jaguar T6 the one that presented the greatest number of descendants ($n = 7$), followed by T5 ($n = 5$). Thus, the genetic contribution of these founders ranges between 0.5 and 3.5%.

The current and potential allelic retention was calculated. Regarding the current allelic retention, it can be observed that this depends mainly on whether the founders have produced descendants and if there has been a genetic contribution from them. It can be observed, in the case of the potential allelic retention, the capacity of allelic retention of the population with the genes coming from those animals that still have not reproduced themselves within the population. (T33, T32, and T20). It can also be observed that individuals T1 and T2, although having contributed to the allelic retention, do not contribute to the potential allelic retention, due to the fact that they have died.

The average value of the current genetic diversity, of the population (GD = 0.7832), was obtained. Similarly, the value of the potential genetic diversity (GDp = 0.9113) was calculated, which is determined when conducting control on the production of cubs in the population, and if there is control over which genes are transmitted to the descendants by means of programs of the management of reproduction toward the future. In this way, it was possible to determine the variations of genetic diversity, during a span of time which permits appreciating that between 1995 and 1996, the genetic diversity of the population increased from 50 to 75%, due to a birth and other reproductive events that took place at that time. A slight increase from the years 1999–2002 can also be appreciated. A genetic value (GV) of 0.7846 was obtained, as was also obtained the value of the equivalent for the current founding genome (FGE = 2.31) and the potential (FGEp = 5.64). In the same way, the value of the equivalent of the current surviving founding genome (FGS = 3.14) and potential (FGSp = 5.64) was obtained. In the graph, an increment of FGE between the years 1995 and 2002 can be observed.

The average coefficient of *F* inbreeding [24] for the population was calculated through time, and it was found on average an $F = 0.0179$, being this the descent of heterozygosity observed, relative to the heterozygosity expected from the founding population. It was found that the coefficient of inbreeding was different from 0, starting in the year 1998.

The size of the effective population or the effective demographic number was also calculated (Ne) based on the history of the population; the current demographic number (*N*) points out the number of live animals within the population, including males as well as females. The primary use of *N* predicts future genetic changes, and its first estimate is based on the amount of diversity of lost genes through generations of cubs in captivity. The second estimate of the Ne is simply based on the number of live males and live females who have produced the current offspring. The proportion of Ne and *N* is employed during the addressing phase of the population, estimating the size of the population that will be needed to achieve genetic goals [19]. Considering the earlier, for the first estimate, the following was found: Ne = 5.26 over the last 1.39 generations, and for the second estimate: Ne = 2.667 for 1.0 male for each 2.0 breeding females; as with the relationship Ne/*N* = 0.2667 and the average value of MK was 0.217.

3.2. Genetic analysis with microsatellites

3.2.1. Genotypes

The genotypes of the jaguars sampled were stabilized from the analysis of the chromatograms obtained after the capillary electrophoresis. It could be determined that the microsatellite, which amplified in fewer individuals, was FCA008 showing the highest index of presentation of null alleles (+0.16). For the 20 jaguars analyzed and the 9 microsatellites, an average of alleles of 5.67 per locus was found, finding the highest number of alleles for the locus FCA 224 (eight alleles) and the lowest for the FCA736 (two alleles). In the wild Colombian jaguar sample for 156 individuals and 12 microsatellites, the average allele number was 13 ± 2.523, being this value significantly higher than that estimated in the captive jaguar population. *Fca136* yielded 18 alleles, whereas *Fca24* and *Fca45* presented 10 alleles.

The average gene diversity (expected heterozygosity) found for the Colombian captive jaguar population was $H = 0.684 \pm 0.230$ by means of microsatellites. It is noteworthy to mention that this average gene diversity value was significantly lower to that estimated in the Colombian wild jaguar population ($H = 0.867 \pm 0.0588$; $t = 2.86$, $p < 0.02$). Identically, the value obtained for the Colombian captive jaguar population was also significantly lower from the wild jaguar populations of Peru and Bolivia ($H = 0.883 \pm 0.045$ and $H = 0.883 \pm 0.043$, respectively; [14]). Thus, the captive jaguar population of Colombia retained about 78% of the genetic diversity found in the north-western South-American wild jaguar population.

In the captive population, FCA 391 and +FCA 220 presented the highest values of heterozygosity and information of polymorphic content (PIC) and the locus with least heterozygosity and PIC was the FCA736. In the wild sample, *Fca176* and *Fca136* were the two markers with the highest values of expected heterozygosity and PIC, meanwhile *Fca08* and *Fca45* were those with the lowest values of gene diversity and PIC. As for the probability of the presentation of null alleles, the highest probability was found for the locus FCA 008. The values of the heterozygous and the homozygous alleles were found for each locus, as well as the frequency of presentation including and excluding the probability of presentation of null alleles. For microsatellite Fca 075, seven (7) alleles were found; being el alelo de 114 pares de bases (bp) the most frequent. For microsatellite Fca 44, five (5) alleles were found; being 146 bp the most frequent. For microsatellite Fca 220, six (6) alleles were found; being 212 the most frequent. For microsatellite Fca 043, three (3) alleles were found, being the 114 bp. This microsatellite was also genotyped for the wild sample showing 13 alleles. For the microsatellite Fca 224, eight (8) alleles were found; being 152 bp the most frequent. For microsatellite Fca 736, two (2) alleles were found; being 125 the most frequent. For microsatellite Fca 008, seven (7) alleles were found; being 110 bp the most frequent. This microsatellite was also genotyped for the wild sample showing 15 alleles. For microsatellite Fca 096, seven (7) alleles were found; being 189 bp the most frequent. This microsatellite was also genotyped for the wild sample showing 14 alleles. For microsatellite Fca 391, six (6) alleles were found, 208 bp being the most frequent. This microsatellite was also genotyped for the wild sample showing 15 alleles. Therefore, in those microsatellites simultaneously genotyped in both jaguar samples, in all the cases the wild jaguar sample yielded a significantly higher number of alleles.

3.2.2. The Hardy-Weinberg equilibrium

In this study, the test of probabilities for each locus was conducted; finding excess of homozygotes for the loci FCA 441, FCA 008, FCA 224, and FCA096 through the estimates F_{is} de Weir & Cockerham's (W&C) [29], and Robertson and Hill's (R&H) [30]; nevertheless, the Fisher method resulted in not being significant for the excess of homozygotes (χ^2: 18.9872 GL: 18.0000, $p = 0.3926$).

3.2.3. Endogamy

When assuming the Hardy-Weinberg equilibrium, the demographic F_{is} statistic was calculated. An excess of heterozygotes was found for this population in the loci FCA075, FCA043, FCA736, FCA220, and FCA391 through the estimates of F_{is} of Weir & Cockerham's (1984) (W&C), and Robertson & Hill's (1984) (R&H); likewise, the result obtained for the F_{is} average for the population was (F_{is} W&C= −0.0157) and (F_{is} R&H = −0.0109). Contrarily to that found in the Colombian captive jaguar population, the Colombian wild jaguar sample showed a significant positive value of F_{is} ($F_{is} = 0.344 \pm 0.042$), which agrees quite well with a large genetic subdivision in this population than in the captive jaguar population of this South America country.

3.2.4. Disequilibrium by linkage

No significant differences for the de test of disequilibrium by linkage were found, which means that each locus, being independent from the others, can be employed as determiners of the genetic diversity of the population. The same was found for the 12 microsatellites applied in the wild jaguar sample.

3.2.5. Pedigree

3.2.5.1. Maternity

The analysis of verisimilitude obtained in CERVUS resulted with assignments of maternity with no knowledge of the father by 34% for the strict confidence interval at 95% (LOD critical = 4) and with a 100% for the assignments for the flexible interval at 85%. The assignments of maternity with knowledge of the father were of 84% for the strict confidence interval at 95% (LOD = 2.05). Only the father (T5) of individuals T7, T8, and the father (T3) of individual T10 were known. Assignments are conducted according to the coupled LOD scores.

3.2.5.2. Paternity

The analysis of verisimilitude obtained in CERVUS resulted with assignments of paternity with no knowledge of the mother of 32% for the strict confidence interval at 95% (LOD critical = 4) and with 100% of assignments for the flexible interval at 85%. Assignments of paternity with knowledge of the mother were of 81% for the strict confidence interval at 95% (LOD = 2.05). Only the mother (T6) of individuals T7, T8, T10 was known. The assignments were conducted according to the coupled LOD scores.

3.2.5.3. Kinship

The probabilities of relationship and kinship assigned by KINSHIP were obtained to a prove H_1 Rp = 0.5 and Rm = 0.5 versus H_0 Rp = 0 and Rm = 0 at a significant level of $\alpha = 0.05, 0.01$, and 0.001, which means that when finding significant and highly significant relationships, it is accepted H_1 where Rp and Rm = 0.5 are the probability that two individuals present a first-degree relationship, be it father and/or mother of son/daughter or brother-sister/-sister/brother. In the results obtained in the simulations, they found significant relationships ($p < 0.05$) in 6 couples of jaguars, highly significant relationships in 3 couples ($p < 0.01$), and significant relationships ($p < 0.001$) for 11 couples.

Taking into account the information of the records of the Studbook, the assignation of paternities and maternities for the CERVUS program and the relationships of kinship obtained by KINSHIP, the construction of the pedigree was conducted, for the population of captive jaguars in the zoological parks of Colombia (**Figure 1**). When conducting the construction of the pedigree, a great association was found, between the animals present at the Matecaña and the Santafe zoological parks. A high genetic relation among the jaguars present at the Santacruz zoological park could be determined, and additionally there were connections in the records of the Studbook between this zoo and the Piscilago, Barranquilla, and Jaime Duque. As for the *fundacion zoologico de Cali*, relationships were found from the records in its historical collections. Regarding the jaguars that are in the Ocarros, no associations were found with other individuals of the captive population in Colombia.

To sum it up, through the CERVUS program, four maternities (T15, T25, and T19 in two occasions) and five paternities (T17, T26, T3, and T16 in two occasions) were assigned. They were assigned by CERVUS as well as by KINSHIP four paternities ($p < 0.001$) (T17, T18, and T5 in

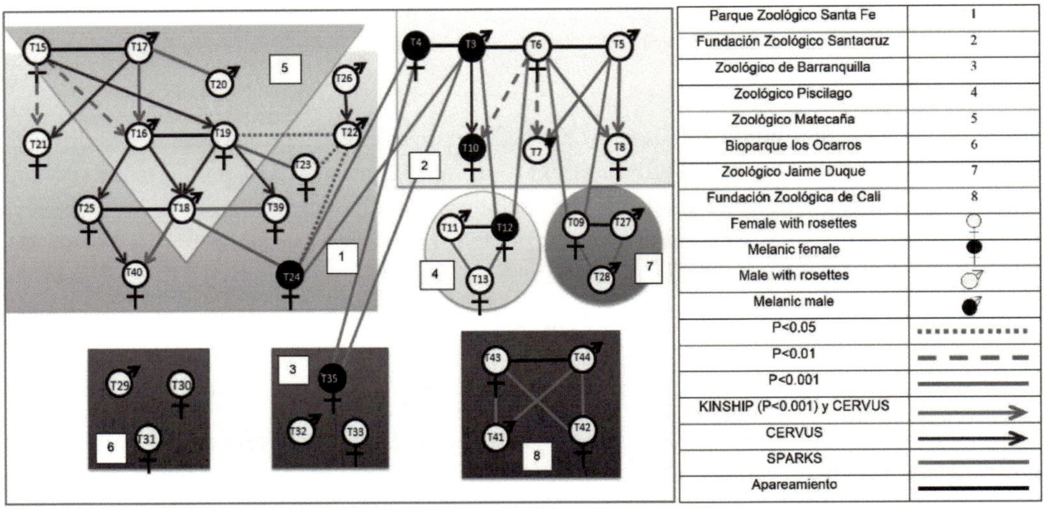

Figure 1. Pedigree was of the population of captive jaguars in the zoological parks of Colombia.

two occasions) and one maternity (T6). In this manner, four maternities were assigned by KINSHIP ($p < 0.01$) for two individuals (T15 in two occasions and T6 two occasions). A significant ($p < 0.05$) relationship was found through KINSHIP, between jaguar T17 and jaguars T23, T19, and T24.

In **Figure 2**, it is possible to observe as an example a paternity and maternity test in jaguars (*P. onca*), employing nine microsatellites, assigned by CERVUS with a significant relationship ($p < 0.001$) found in the KINSHIP program, between father and son and significant ($p < 0.001$) between mother and son. The results are presented graphically through the chromatograms of each system of microsatellites employed and numerically highlighting in grey the size of the alleles inherited.

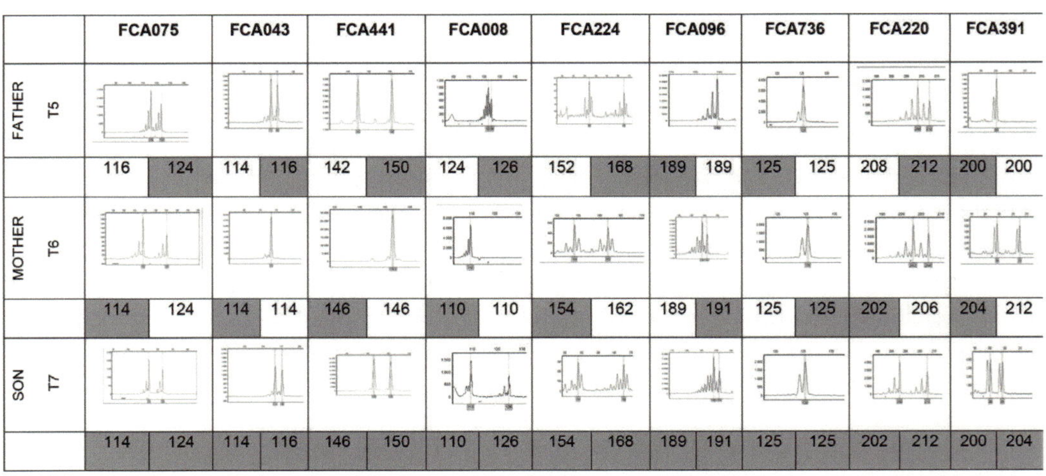

Figure 2. Paternity and maternity of a triplet of captive jaguars.

4. Discussion

In this study, it could be proven that the population of captive jaguars in zoological parks in Colombia is a population that presents a high genetic variability and that is a population with the necessary conditions to be an important model as a genetic reservoir to revitalize natural populations given the case that they might be isolated and present low levels of genetic diversity. Nevertheless, this study also shows that the Colombian zoological captive jaguar population only has a fraction of the overall genetic diversity found in the wild jaguar population of this South-American country [7, 13, 14, 31].

As for the genetic analysis from the regional Jaguar Studbook, it could be determined that regarding the mean kinship (MK) on average (MK = 0.217), the values found for the captive population in Colombia are less than the MK reported by Drury [32] for captive jaguars in Europe (MK = 0.238), which offers a large indicative as future potential to reduce and/or events of inbreeding. It was found that there exists representation and genetic contribution of founders from wildlife in the current population due to breeding in captivity.

Regarding the genetic diversity according to the records found in the regional Studbook, a clear tendency was found, of an increase that went from 50% in the year 1994 to 72% in the year 1996 and reaching nearly 78% in the year 2002, due to the reproductive processes and the births that took place in this lapse of time.

As for the genetic diversity in the analysis of microsatellites, it was found that it is in the order of 68%. Such a result is found in the levels reported by Eizirik et al. [12], where he studied levels of genetic diversity between 62 and 73%. This last study analyzed a large fraction of jaguars living in the periphery of the geographical distribution of this species (Mexico and Central America and Southern Brazil and Argentina). However, these gene diversities were considerably minor than those found by Ruiz-García et al. [7, 13], who reported a gene diversity of 84%, or the gene diversity showed in this study and Ruiz-García et al. [14] for a larger Colombian wild jaguar sample around 87% [14]. Henceforth, the Colombian captive jaguar sample contained around 78% of the gene diversity found in wild conditions. Due to this, it is possible to determine that the genetic diversity of captive jaguars in Colombia is at a high level; nevertheless, it is necessary to follow the alignments of reproductive population management for the species and thus reach a level close to 90% and maintain it for a 100-year period as it is proposed in the plans of survival of species [33].

As for the analysis conducted of the microsatellites, a smaller number of alleles were found, for locus FCA96 (7 alleles), FCA08 (7 alleles), and FCA391 (6 alleles), than the number reported by Ruiz-García et al. [34] where the number of alleles found was greater for FCA96 (15 alleles), FCA08 (15 alleles), and FCA391 (13 alleles) [34], and by Ruiz-García et al. [13], where it was found for Fca96 (13 alleles), Fca08 (13 alleles), and Fca391 (13 alleles) [13]. Regarding locus Fca96, alleles of 185pb 187pb, and 189pb have been reported in Amazonian jaguars of the subspecies (*P. onca onca*) and of 183pb in Guatemala of the subspecies (*P. onca goldmani*) [13], which in this investigation were found with the following allelic frequency: 185 pb = 0.05, 187pb = 0.175, 189 pb = 0.35 and for the 183 pb = 0.075, found only in the T15, T16, and T19 at the Matecaña zoo. As for loci T15, T16, and T19 located at Matecaña zoo, an allele of 205 pb was reported only in jaguars of Peru [13] found in this study in eight specimens with a high allelic frequency of 0.21. Regarding locus Fca008, an allele of 124pb was reported, only in jaguars from the central Amazonia [13], found through this study, two specimens with a high allelic frequency of 0.09 in specimens T5 and T8 located at *Fundación Zoológico Santacruz*. This could mean that the Colombian captive jaguar population could be composed by animals proceeding to three different putative subspecies (although the subspecies concept applied to the jaguar is discussed by Ruiz-García et al. [14]), with greater predominance of the Amazonian jauar *P. onca onca* [35], followed by the Peruvian jaguar *P. onca ucayale* [36] and lastly by *P. onca centralis*, not ruling out the *P. onca goldmani*, having found alleles of this subspecies.

In this study, it was found that for the loci analyzed, there exists Hardy-Weinberg equilibrium, which contrasts with the findings of Ruiz-García et al. [13, 31] who reported the inexistence of the Hardy-Weinberg equilibrium for excess of homozygotes which was attributed to an effect of the demographic subdivision or Wahlund effect in the total wild jaguar population of Colombia and north-western South America. On the other hand, Eizirik [12] reported deviations from the Hardy-Weinberg equilibrium in 6 loci of the 29 analyzed from the population of jaguars studied. Thus, the estimated equilibrium Hardy-Weinberg is a point that needs greater approach in order to clarify and determine if the disequilibrium found by Ruiz-García is due to the effect of demographic subdivisions, or Wahlund effect, or if this finding is a dependent reflection of loci analyzed. However, notwithstanding that in this study we find the existence of equilibrium for the loci analyzed, the size of the sample is reduced; therefore, inference could not be extensively extrapolated to the populations of jaguars in the wild. In this way, it would be important to continue this kind of analysis in future investigations.

The results of the genetic analysis of the Regional Jaguar Studbook for Colombia in the PM2000 program clearly show that according to the records regarding inbreeding, the maximum value was found for F in jaguars in Europe ($F = 0.08$), a more superior value to that found in the Jaguar Studbook for Colombia ($F = 0.0179$), event that has only been registered since the year 1998. Nevertheless, posterior to the analysis employing microsatellites, it could be determined through statistic calculation F_{is} as a determiner of inbreeding of the population based on analysis data of DNA, contrasts with the results obtained in the analysis of records for the coefficient of inbreeding F.

As for the value found for F_{is}, it represents that in the population, an excess of heterozygotes is found, which when averaged in all the loci analyzed and being a negative value, it is an indicator that the population is not in endogamy; on the contrary, it is in exogamy (the F_{is} value is negative if the heterozygosity observed is greater than expected, which means that the average observed is greater than the one expected). The inbred populations are characterized by an increment in their homozygosis or a diminishment in their heterozygosis; thus, a negative value of F_{is} signifies that there exists an increment in heterozygosis and that the population is in exogamy [37].

Before conducting the study, the origin of the pedigree population of jaguars was known by 28.9% ($n = 11$) after the construction of the pedigree of the population through the analysis with the microsatellites, 53% ($n = 21$) could be known following an assignation of paternities and maternities through the CERVUS and KINSHIP programs, which proves an optimization of the information in 98%, offering great advantages from the use of microsatellites as a complement in the actualization of the Studbook.

From this, it was possible to obtain a precise register of the origin of the live specimens within the population, which can be employed in the future programs of population management, involving controlled crossings in order to avoid the presence of inbreeding and thus maintain the jaguar population as a variable population, in genetic terms, as a model of investigation.

For some species where populations in the wild are highly diminished, the survival of the species depends on the propagation of captive animals. In these cases, the entire genetic pool of a species represented by the genetic contribution of the captive founders to the following generations of individuals bred and raised in captivity is presented as a closed system. In these cases, the new mutations that occur within the population provide the only new source of genetic variation. This process takes place at an extremely low rate and provides significant changes only at long intervals of time or when populations are extremely large. In the cases where the wild populations are stable in number and distribution so that potentially they can be sampled and could provide immigrants to the captive population, this access to new sources of variation provides the opportunity to preserve larger proportions of wild genetic pool, employing fewer captive individuals in a greater way than in closed populations. It is at this point where starting from the historical knowledge of animal populations is possible to pose concrete solutions with the purpose of conserving all the species that mankind has placed under threat, as is the jaguar. The implementation of programs of genetic analysis in the collections present in captivity contributes to the conservation of the species threatened, through the maintenance of viable populations as genetic reservoirs that can be used periodically to reinforce, revitalize, or reestablish captive populations and when necessary in cases of wild populations with low genetic variability. It also permits the identification of subspecies that even though phenotypically might seem similar, they have different geographical origins and have suffered a particular process of speciation. Although in the case of the jaguar, both morphological and molecular studies put in doubt the existence of subspecies [12–14, 34, 38, 39].

The scope of the instauration of programs of genetic investigation in the zoological parks and centers of rehabilitation of fauna, contributor to the ordering of captive collections, provides clear tools in the face of the movement of specimens from one place to another, in order to guarantee efficient reproductive and viable processes.

In addition, it contributes to the liberation of animals after the processes of rehabilitation in zones corresponding to the natural genetic origin of their congeners. The implementation of a genetic analysis of a population in captivity requires systematic steps in order to obtain the greatest quantity of possible information, such as the construction of a "pedigree" in the population, the identification of founding parental, the genetic contribution of each parental to the population, the estimate of the loss of alleles due to endogamy or "bottleneck" effects, and estimates of coefficients of inbreeding and making of tests employing molecular markers. In this way, adequate genetic techniques can be implemented to clarify the blanks of information about unknown specimens.

Therefore, this was the first study of records of captive jaguars in Colombia where it was possible to obtain relevant information regarding the identification of individual's founders of the population and their descendants. This kind of studies could be carried out in other Latin American countries, where the captive populations of jaguars are large (for instance, Peru, Bolivia, and Brazil), and thus these captive populations could be decisive to the conservation of the largest wild cat of the Americas.

Acknowledgements

The authors would like to express their acknowledgments to the Colombian Association of Zoological Parks and Aquariums (ACOPAZOA): the Zoológico Santacruz Foundation: Dr. Angelica Moreno, John Vidal. Zoológico Santafé: Dr. Marta Ocampo. Zoológico Matecaña: Dr. Delio Orjuela, Sandra Correa.Zoológico Piscilago: Dr. Ana Carolina Falla. Fundación zoológica de Cali: Dr. Jorge Gardeazabal, Dr. Juliana Peña. International Species Information System ISIS: Laurie Bingaman. Venezuelan Institute of Scientific Investigations (IVIC): Dr. Kate Rodríguez Clark. Institute of Genetics of the National University: Dr. Victor Vera, Dr. Luis Fernando Cadavid, Miguel Novoa, Dr. Fernando Ariza. Dr. Paul Bloor. Dr. Carolina Corrales. National Zoo, Smithsonian Institution: Dr. Christine Bozarth, Dr. Joseph Kolowski, Dr. Francisco Dallmeier. National Cancer Institute: Dr. Stephen J. O'Brien, Dr. Warren Johnson, Victor David, Bailey Kessing. Maria Clara Espinel, Consuelo Espinel, Julia E González.

Author details

Santiago Jimenez Gonzalez[1*], Manuel Ruiz García[2], Jesús Maldonado[3], Ligia Jimenez[1], Alberto Sánchez[1], Tammy Wilbert[3], Susan L. Castro[1], Jimmy H. Vargas[1], Sonia R. Quintanilla[2], Haydy Monsalve[4] and Claudia Jimenez Escobar[1]

*Address all correspondence to: santiaklen@gmail.com

1 National University of Colombia, Bogotá, Colombia

2 Pontificia Universidad Javeriana, Bogotá, Colombia

3 National Zoo, Smithsonian Institution, Washington D.C., USA

4 Fundación Zoológico Santacruz, Cundinamarca, Colombia

References

[1] Weber JL, May PE. Abundant class of human DNA polymorphisms which can be typed using the polymerase chain reaction. American Journal of Human Genetics. 1989;**44**(3):388-396

[2] Menotti-Raymond M, O'Brien SJ. Dating the genetic bottleneck of the African cheetah. Proceedings of the National Academy of Sciences United States of America. 1993;**90**(8):3172-3176

[3] Menotti-Raymond MA, O'Brien SJ. Evolutionary conservation of ten microsatellite loci in four species of Felidae. Journal of Hereditary. 1995;**86**(4):319-322

[4] Culver M, Menotti-Raymond MA, O'Brien SJ. Patterns of size homoplasy at 10 microsatellite loci in pumas (*Puma concolor*). Molecular Biology and Evolution. 2001;**18**(6):1151-1156

[5] Buckley-Beason VA, et al. Molecular evidence for species-level distinctions in clouded leopards. Current Biology. 2006;**16**(23):2371-2376

[6] Luo SJ, et al. Subspecies genetic assignments of worldwide captive tigers increase conservation value of captive populations. Current Biology. 2008;**18**(8):592-596

[7] Ruiz-García MM, et al. Genética de poblaciones amazónicas: la historia evolutiva del jaguar, ocelote, delfín rosado, mono lanudo y piurí, reconstruida a partir de sus genes. Animal Biodiversity and Conservation. 2007;**30**(**2**):115-130

[8] Ruíz-García M, et al. Craneometric and microsatellite genetic differentiation among putative ocelot subspecies (*Leopardus pardalis*) throughout Latin America. Molecular Population Genetics, Evolutionary Biology and Biological Conservation of the Neotropical Carnivores. New York (USA): Nova Science Publishers Inc.; 2013

[9] Cossíos D, et al. Influence of ancient glacial periods on the Andean fauna: The case of the pampas cat (*Leopardus colocolo*). BMC Evolutionary Biology. 2009;**9**(1):68

[10] Cossíos ED, et al. Population structure and conservation of a high-altitude specialist, the Andean cat *Leopardus jacobita*. Endangered Species Research. 2012;**16**(3):283-294

[11] Ruiz-García M, et al. Population genetics and spatial structure in two Andean cats (the Pampas cat, *Leopardus pajeros* and the Andean Mountain Cat, *L. jacobita*) by means of nuclear and mitochondrial markers and some notes on skull biometrics. Molecular Population Genetics, Evolutionary Biology, and Biological Conservation of Neotropical Carnivores. [Hauppauge] New York, NY: Nova Publishers; 2013

[12] Eizirik E, et al. Phylogeography, population history and conservation genetics of jaguars (*Panthera onca*, Mammalia, Felidae). Molecular Ecology. 2001;**10**(1):65-79

[13] Ruiz-García M, et al. DNA microsatellite characterization of the jaguar (*Panthera onca*) in Colombia. Genes & Genetic System. 2006;**81**(2):115-127

[14] Ruiz-García M, et al. Population genetics and phylogeography of the largest wild cat in the Americas: An analysis of the jaguar by means of microsatellites and mitochondrial gene sequences. Chapter 11. In: Molecular Population Genetics, Evolutionary Biology and Biological Conservation of the Neotropical Carnivores. New York (USA): Nova Science Publishers., Inc.; 2013. pp. 413-464

[15] Driscoll C, et al. Genomic microsatellites as evolutionary chronometers: A test in wild cats. Genome Research. 2002;**12**(3):414-423

[16] Ruíz-García M, et al. Determination of microsatellite DNA mutation rates, mutation models and mutation bias in four main Felidae lineages (European wild cat, *F. silvestris*; ocelot, *Leopardus pardalis*; puma, *Puma concolor*; jaguar, *Panthera onca*). Chapter 10. In: Molecular Population Genetics, Evolutionary Biology &; Biological Conservation of Neotropical Carnivores. New York (USA): Nova Science Publishers Inc.; 2013

[17] Jimenez Gonzalez S, et al. Reproductive analysis of male and female captive jaguars (*Panthera onca*) in a Colombian zoological park. Theriogenology. 2017;**89**:192-200

[18] Jiménez-González S, Moreno WF. Manejo y anestesia en cinco especies de félidos en cautiverio en parques zoológicos de Colombia. Revista UDCA, Actualidad y Divulgación Científica. 2004;**6**: pp. 73-83.

[19] Pollak JP, Lacy RC, Ballou JD. Population Management 2000, version 1.213. Brookfield, IL: Chicago Zoological Society; 2005

[20] Nei M. Analysis of gene diversity in subdivided populations. Proceedings of the National Academy of Sciences United States of America. 1973;**70**(12):3321-3323

[21] Willis K, Wiese RJ. Effect of new founders on retention of gene diversity in captive populations: A formalization of the nucleus population concept. Zoo Biology. 1993;**12**(6):535-548

[22] Wright S. Evolution and the Genetics of Populations, Volume 2: Theory of Gene Frequencies. Chicago, IL: University of Chicago Press; 1969

[23] Archie JW. Statistical analysis of heterozygosity data: Independent sample comparisons. Evolution. 1985;**39**(3):623-637

[24] Wright S. Coefficients of inbreeding and relationship. The American Naturalist. 1922; **56**(645):330-338

[25] Lacy RC. Clarification of genetic terms and their use in the management of captive populations. Zoo Biology. 1995;**14**(6):565-577

[26] Raymond M. GENEPOP (version 1.2.): Population genetics software for exact tests and ecumenicism. Journal of Heredity. 1995;**86**:248-249

[27] Rousset F. Genepop'007: A complete re-implementation of the genepop software for Windows and Linux. Molecular Ecology Resources. 2008;**8**(1):103-106

[28] Goodnight KF, et al. Kinship 1.2. User Manual. Houston, TX: Rice University. Dept. of Ecology & Evolutionary Biology; 2004

[29] Weir BS, Cockerham CC. Estimating F-statistics for the analysis of population structure. Evolution. 1984;**38**(6):1358-1370

[30] Robertson A, Hill WG. Deviations from Hardy-Weinberg proportions: Sampling variances and use in estimation of inbreeding coefficients. Genetics. 1984;**107**(4):703-718

[31] Ruíz García M, et al. Genética molecular de poblaciones aplicada al estudio de dos grandes carnívoros neotropicales (Tremarctos ornatus Cuvier, 1825 – Oso andino y *Panthera onca* linné, 1978 – Jaguar). Boletín de la Real Sociedad Española de Historia Natural. 2003

[32] Drury H, H.S.S.I. Investigation into inbreeding en European Zoo Population of jaguars (Panthera onca). Zoo Research News. BIAZA Research Newsletter. 2005;**6**(4)

[33] Ballou JD, Foose T. Demographic and genetic management of captive populations. Wild Mammals in Captivity, Principles and Techniques. Chicago, IL: University of Chicago Press; 1996

[34] Ruiz-García M. Estudios moleculares (microsatélites) en el caso de Primates y Félidos neotropicales comportan una nueva perspectiva. Diversidad genética como herramienta de zonificación ambiental. Bogotá DC: Instituto Amazónico de Investigaciones, Imani and Instituto de Ciencias Naturales, Universidad Nacional de Colombia; 2001

[35] Pocock RI. The races of the jaguar, *Panthera onca*. Novitates Zoology. 1939;**41**:221-240

[36] Nelson EWGE. Revision of the jaguars. Journal of Mammalogy. 1933;**14**:221-240

[37] Allendorf FW, Luikart G. Conservation and the Genetics of Populations. Oxford, UK: Blackwell Publishing; 2007

[38] Larson SE. Taxonomic re-evaluation of the jaguar. Zoo Biology. 1997;**16**:107-120

[39] Ruiz-García MEP. Craniometric variation in jaguar subspecies (*Panthera onca*) from Colombia. Chapter 12. In: Molecular Population Genetics, Evolutionary Biology and Biological Conservation of the Neotropical Carnivores. New York (USA): Nova Science Publishers Inc.; 2013

Chapter 6

Ecology and Evolution of Melanism in Big Cats: Case Study with Black Leopards and Jaguars

Lucas Gonçalves da Silva

Additional information is available at the end of the chapter

http://dx.doi.org/10.5772/intechopen.69558

Abstract

Variations in animal coloration have intrigued evolutionary biologists for a long time. Among the observed pigmentation polymorphisms, melanism has been reported in multiple organisms (influencing several biological factors), and classical hypothesis has suggested that such variant can present adaptive advantages under certain ecological conditions. In leopards (*Panthera pardus*) and jaguars (*Panthera onca*), melanism is caused by recessive and dominant mutations in the *ASIP* and *MC1R* genes, respectively. This chapter is focused on melanism in these two species, aiming to analyze its geographic pattern. About 623 leopard and 980 jaguar records that were used as baseline for modeling and statistical analyses were obtained. The frequency of melanism was 10% for both species. In leopards, melanism was present in five subspecies and strongly associated with moist forests, especially in Southeast Asia. In jaguars, melanism was totally absent from open and periodically flooded landscapes; in contrast, forests displayed a frequency that was similar to the expectations. The analyses of the environmental predictors suggest a relevant role for factors such as moisture and temperature. These observations support the hypothesis that melanism in big cats is not a neutral polymorphism (influenced by natural selection), leading to a nonrandom geographic distribution of this coloration phenotype.

Keywords: natural selection, geographic distribution, black panther, phenotypes, biomes

1. Introduction

The adaptive relevance of animal coloration has been explored and discussed for over a century (e.g., [1–4]). Pigmentation phenotypes have often been inferred to present adaptive roles in ecological, physiological, and/or behavioral processes such as camouflage, intra- and interspecific communication, and thermoregulation [3–6]. The broad phenotypic variation of vertebrates observed in nature is one of the questions that have long intrigued evolutionary biologists, including its adaptive significance and genetic basis. Despite the interest in the subject, relatively few studies have addressed the association between phenotypes in natural populations and the environments in which organisms occur, aiming to investigate evolutionary processes involved in the generation and maintenance of coloration diversity and the environmental characteristics that influence the adaptive significance of phenotypes [7–9].

Melanism is a color polymorphism that is common in various groups of organisms, in which the skin/fur/plumage is darker than what would be considered the normal or "wild" phenotype. There are classical hypotheses that postulate an adaptive role of melanism in different species, involving many potential impacts on survival or reproduction [4–6, 10]. Several biological factors such as thermoregulation, susceptibility, or response to disease, camouflage, aposematism, sexual selection, and reproductive success could be directly influenced by melanism [5].

The occurrence of melanism is rather common in Felidae, having been documented in 13 of the 38 felid species, evolved independently at least eight times within the family [11–13], in some cases reaching very high frequencies in natural populations [14, 15]. In none of them has it reached fixation, but rather always exists as a polymorphic phenotype, and it is present only in two species of *Panthera*: the leopard (*P. pardus*) and the jaguar (*P. onca*) in contrast of the spotted wild phenotype present in these species. These observations support the hypothesis that melanism can provide an adaptive advantage in certain ecological conditions [6, 16], and exists the hypothesis of an association between darker individuals and wetter areas with dense vegetation (e.g., tropical forests) [1, 4, 10, 17]. In addition, there have also been suggestions of the potential for negative selection against dark individuals in open areas where the sunlight/radiation levels and mean temperatures are high [5, 10, 17, 18]. These hypotheses have been commonly mentioned in the technical literature as anecdotal postulates and also appeared in the popular culture for some time.

The leopard is the largest spotted cat in Africa and Asia [19] and an important extant representative of genus *Panthera* [19, 20]. Its historical distribution is the broadest among all felids (from the Russian far east to Africa), encompassing a diverse array of habitats, from deserts to rainforests, and from the humid tropics to temperate zones [19]. In contrast, the jaguar (*Panthera onca*) is the largest wild cat in the Americas, and the only extant representative of genus *Panthera* in the New World [19]. Its current distribution stretches over 18 nations, from the southern United States to Argentina [21].

Although variant pigmentation phenotypes in vertebrates are caused by several genes [5], in the case of leopards it has been shown that melanism is induced by a recessively inherited mutation in the *ASIP* (*agouti signaling protein*) gene, which leads to a nonsense mutation predicted completely ablate *ASIP* function and thus induce black pigmentation [12]. In other hand, melanism in jaguars is inherited as a dominant trait, caused by a 15-base-pair deletion in the *MC1R* gene that leads to a "gain of function" mutation favoring the production of dark melanin (eumelanin) in the background regions of the coat [11, 22]. Although the trait has been well known in these species for many years and easily identifiable in nature (e.g., [15, 17, 23–26]), its

http://dx.doi.org/10.5772/intechopen.69558

geographic distributions, as well as the environmental factors that may influence its persistence in natural populations, are being discovered recently [26–29].

This chapter presents the geographic distribution of melanism in *P. pardus* and *P. onca* in nature, as well as the response to environmental predictors aiming to evaluate the adaptive relevance of this phenotype. There were considered and tested two alternative hypotheses: (1) melanism present throughout the species' distribution, occurring randomly across all environments (i.e., the absence of association with different landscape conformations) and (2) melanism distributed according to biogeographic constraints.

2. Methods

2.1. Species database

The data set comprised location records points from the entire historical range of leopards (**Figure 1**) and jaguars (**Figure 2**), encompassing various biomes (e.g., moist and dry forests, grasslands and desert areas). These records were obtained from five different sources: (1)

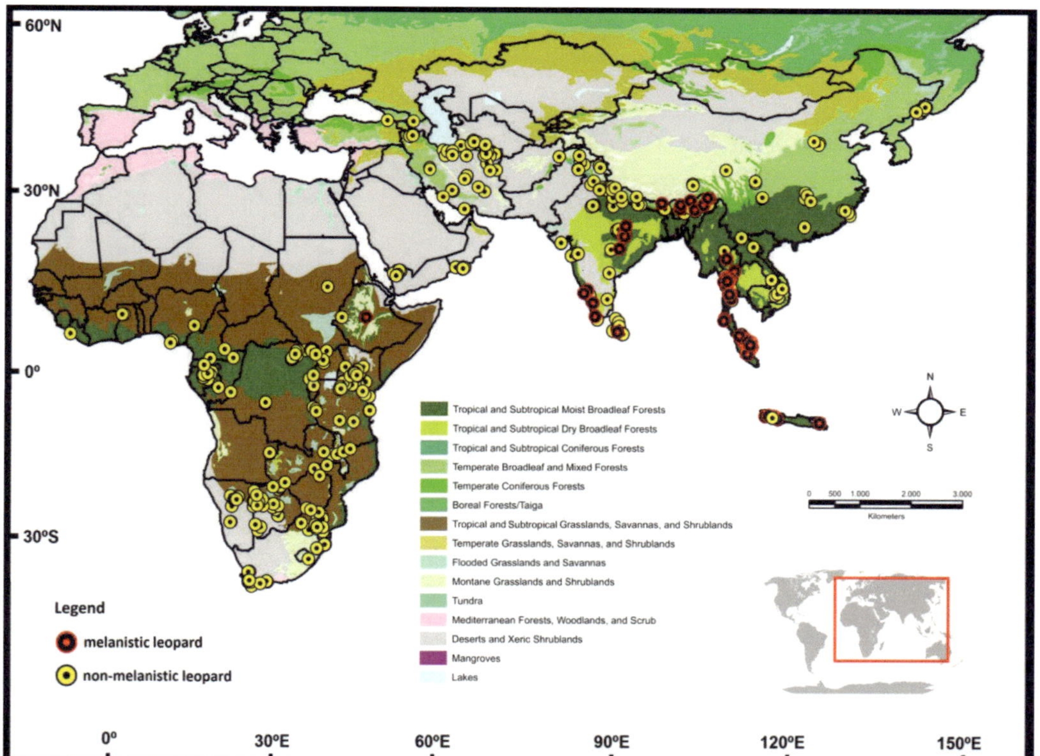

Figure 1. Location of melanistic and nonmelanistic leopard records, overlaid on a map of terrestrial biomes (based on Ref. [30]).

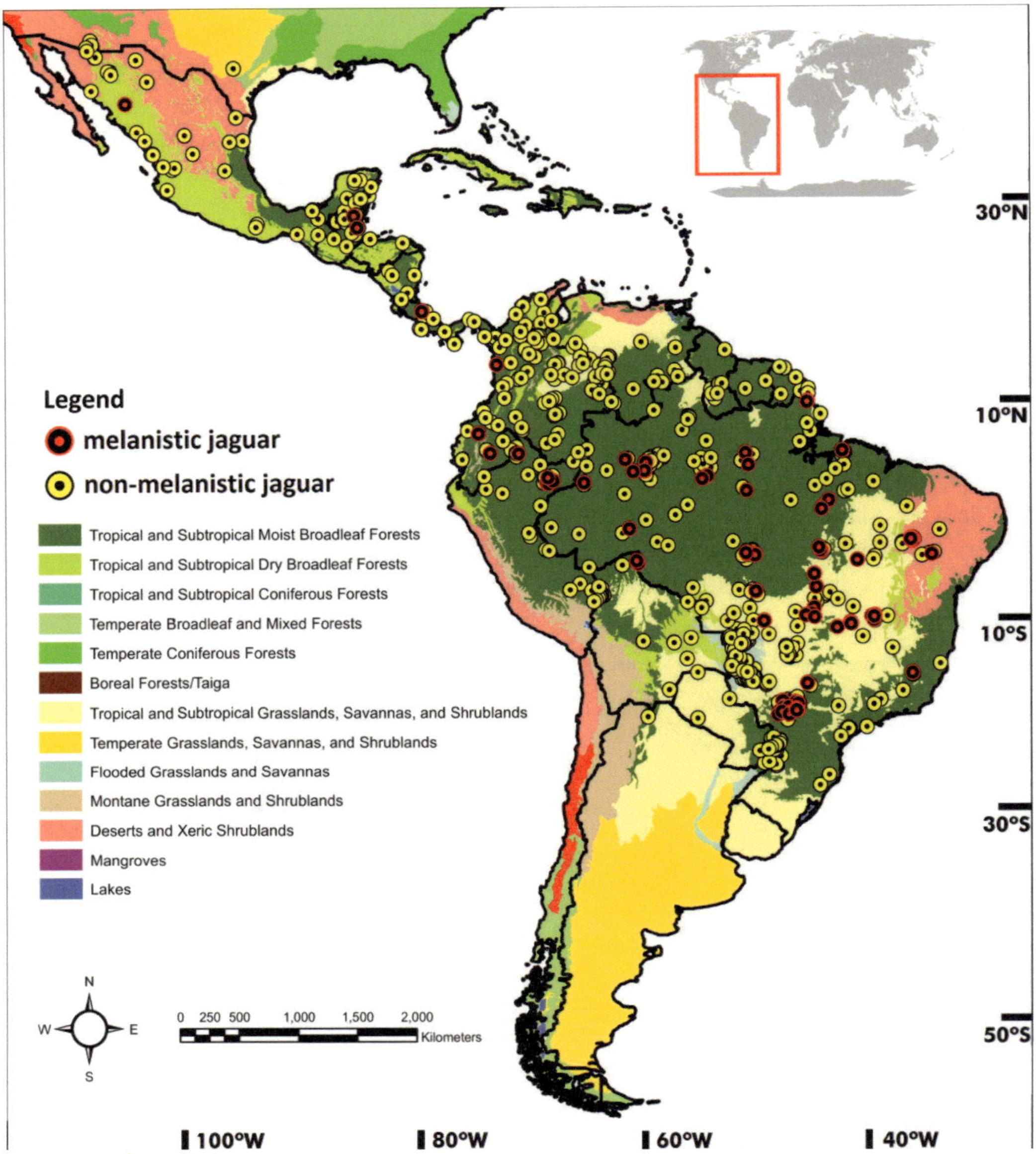

Figure 2. Location of melanistic and nonmelanistic jaguar records, overlaid on a map of terrestrial biomes (based on Ref. [30]).

individuals kept in scientific collections that possessed information on the geographic coordinates of the sampling location as well as on coat color (or, preferably, available skin for direct assessment and photographic documentation of coat color); (2) individuals captured or found dead during field ecology studies; (3) field-collected fecal samples whose melanism status could be confidently inferred using a molecular assay (following [22]); (4) camera trap records; and (5) samples available in online databases with precise geographic origin and available source information.

Location records of individuals confirmed to be melanistic or nonmelanistic were used in the statistical analyses based on the frequency of these phenotypes for both species. As there was no type of bias in our records with respect to coloration phenotype (i.e., sampling was random with regard to coat color), it was assumed that the frequency in which melanism appears in the total data set represents the overall frequency in the species, which provided a null hypothesis against which we tested potential deviations in different regions.

2.2. Environmental predictors

The occurrence of different phenotypes throughout leopard and jaguar distributions was mapped by building a database of location records of melanistic and nonmelanistic individuals. All the records were converted into degree coordinates, using the *WGS84* standard reference system. Additionally, biome and terrestrial ecoregion shapefiles [30] were used as mask layers to extract and analyze information about natural landscapes in which the phenotypes occur.

For modeling the potential distribution of melanism, in the initial analysis we considered 37 explanatory environmental predictors and landscape data that covered 100% of leopards and jaguars known distributions International Union for Conservation of Nature (IUCN) data. We used 35 environmental variables obtained from the Worldclim (http://www.worldclim.org) and Climond (http://www.climond.org) databases: annual mean temperature (Bio01), mean diurnal temperature range (Bio02), isothermality (Bio03), temperature seasonality (Bio04), max temperature of warmest week (Bio05), min temperature of coldest week (Bio06), temperature annual range (Bio07), mean temperature of wettest quarter (Bio08), mean temperature of driest quarter (Bio09), mean temperature of warmest quarter (Bio10), mean temperature of coldest quarter (Bio11), annual precipitation (Bio12), precipitation of wettest week (Bio13), precipitation of driest week (Bio14), precipitation seasonality (Bio15), precipitation of wettest quarter (Bio16), precipitation of driest quarter (Bio17), precipitation of warmest quarter (Bio18), precipitation of coldest quarter (Bio19), annual mean radiation (Bio20), highest weekly radiation (Bio21), lowest weekly radiation (Bio22), radiation seasonality (Bio23), radiation of wettest quarter (Bio24), radiation of driest quarter (Bio25), radiation of warmest quarter (Bio26), radiation of coldest quarter (Bio27), annual mean moisture index (Bio28), highest weekly moisture index (Bio29), lowest weekly moisture index (Bio30), moisture index seasonality (Bio31), mean moisture index of wettest quarter (Bio32), mean moisture index of driest quarter (Bio33), mean moisture index of warmest quarter (Bio34), and mean moisture index of coldest quarter (Bio35). In addition, we included data on altitude (obtained from the SRTM [http://www2.jpl.nasa.gov/srtm]) as well as on landscape surface cover (obtained from ESA GlobCover Project 2009 [http://due.esrin.esa.int/page_globcover.php]). All variables were used at a fine (~1 km) spatial resolution.

Since correlation among explanatory predictors can lead to model overfitting, we computed Pearson's correlation coefficient (r) between each pair of variables [31–33]. The correlation was assessed by extracting predictor information from 10,000 unique and randomly generated points within the present geographic distribution layer of both species (obtained from IUCN and complemented with our own database records). Then, the predictors that were not highly correlated to each other were selected, using $r = 0.7$ as the cutoff value, assuming that they are sufficient for modeling the geographic distribution of these species, and the distribution of melanism within it.

2.3. Modeling procedures and statistical analyses

The models for the overall distribution of leopards and jaguars and the spatial distribution of melanism were generated using the maximum entropy algorithm implemented in the software Maxent [34] along with associated statistical tools [35–38]. The total set of location records divided by each phenotype for both species, 70% of which were used for training and 30% for testing the models, was employed. The data were sampled using the bootstrap routine of 10 random partitions with replacement [39]. All runs were configured in random seed, convergence threshold of 1E-5 with 500 iterations and 10,000 hidden background points [40]. Model performance was assessed by the AUC (area under curve) value for the receiver operating characteristic (ROC curve) and the binomial probability [39, 41], aiming to obtain models of continental-scale distribution of distinct phenotypes.

The statistical analysis of record distribution for each biome was performed using the chi-square test with Fisher correction [42]. The basic approach was to test differences between the observed and expected frequencies of melanism for each different ecoregion. As there was no detected bias with respect to the sampling of different phenotypes at any location (i.e., our sampling was random in regard to coat color), we used the overall frequency of melanism across the whole range to generate the expected number of melanistic records per biome.

3. Results

3.1. Leopards

About 623 leopard samples, comprising 556 nonmelanistic individuals and 67 melanistic individuals (**Figure 1**), providing a broad coverage of the known leopard current range, were obtained. Melanism presented a global frequency of 10.75% across the species' range, with regional frequencies varying among different landscapes (biomes and ecoregions). The confirmed presence of melanistic leopards was recorded only in five of the nine valid subspecies (**Figure 3**): Africa (*P. p. pardus*), Central India, Nepal, and Bhutan (*P. P. fusca*), Sri Lanka (*P. p. kotiya*), Southeast Asia (*P. p. delacouri*), and Java (*P. p. melas*). All of these regions contained new records for areas in which melanism had been previously described, as well as representation of additional areas. Melanism was absent in the leopard subspecies occurring in the Russian Far East (*P. p. orientalis*), Central China (*P. p. japonensis*), Arabian Peninsula (*P. p. nimr*), and Middle East (*P. p. saxicolor*).

Although leopards were found in more than 100 ecoregions, melanism was most common in tropical and subtropical moist forests (59 of 67 records), especially in the Indian Ghats, Javan forests, Kayah-Karen/Tenasserim forests (Southeast Asia), and Peninsular Malaysian rain forests (Southeast Asia). All of these records were consistent with high suitability of occurrence

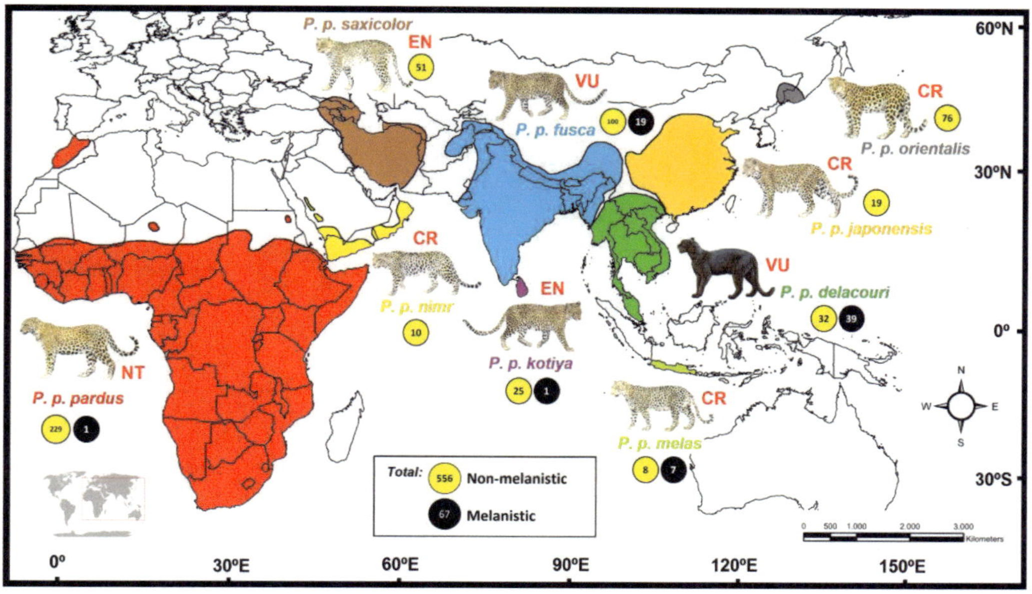

Figure 3. Location records database of leopards, divided by each valid subspecies (based on Uphyrkina et al. [55], with information about IUCN conservation status (NT = near threatened, VU = vulnerable, EN = endangered, and CR = critically endangered).

in the model for melanistic leopards (**Figure 4**). Differences among biomes were significant, especially in tropical and subtropical moist broadleaf forests, where 30% of the animals were black, which is almost three times more than expected if melanism were an evenly distributed neutral polymorphism. In contrast, the frequency of melanism was significantly lower than expected in deserts/xeric shrublands, temperate broadleaf and mixed forests, as well as tropical/subtropical grasslands, savannas, and shrublands.

Niche models were considered satisfactory (AUC $\geq$ 0.9): nonmelanistic model (AUC training = 0.926, test = 0.924) and melanism (AUC training = 0.976, test = 0.963). Control and melanism-predicted distributions generated through niche models are presented in **Figure 4A** and **B**. This assessment allowed a comparison between the overall range of leopards and the presence of melanistic animals, showing regional enrichment for this variant in some areas, as well as its absence in many others. When were analyzed the environmental variables that were most influential on the models, we observed that predictors related to moisture to have the largest effect.

3.2. Jaguars

About 980 jaguar samples, 884 of which were nonmelanistic individuals and 96 were melanistic animals, were obtained. The overall frequency of melanism was 9.80% across the species' range. Intriguingly, almost the same frequency expected for leopards. Most of the records of melanistic animals (92 in total) were located in South America. Moreover, all regions that had previously been reported as potential sites of melanistic jaguar occurrence in different biomes

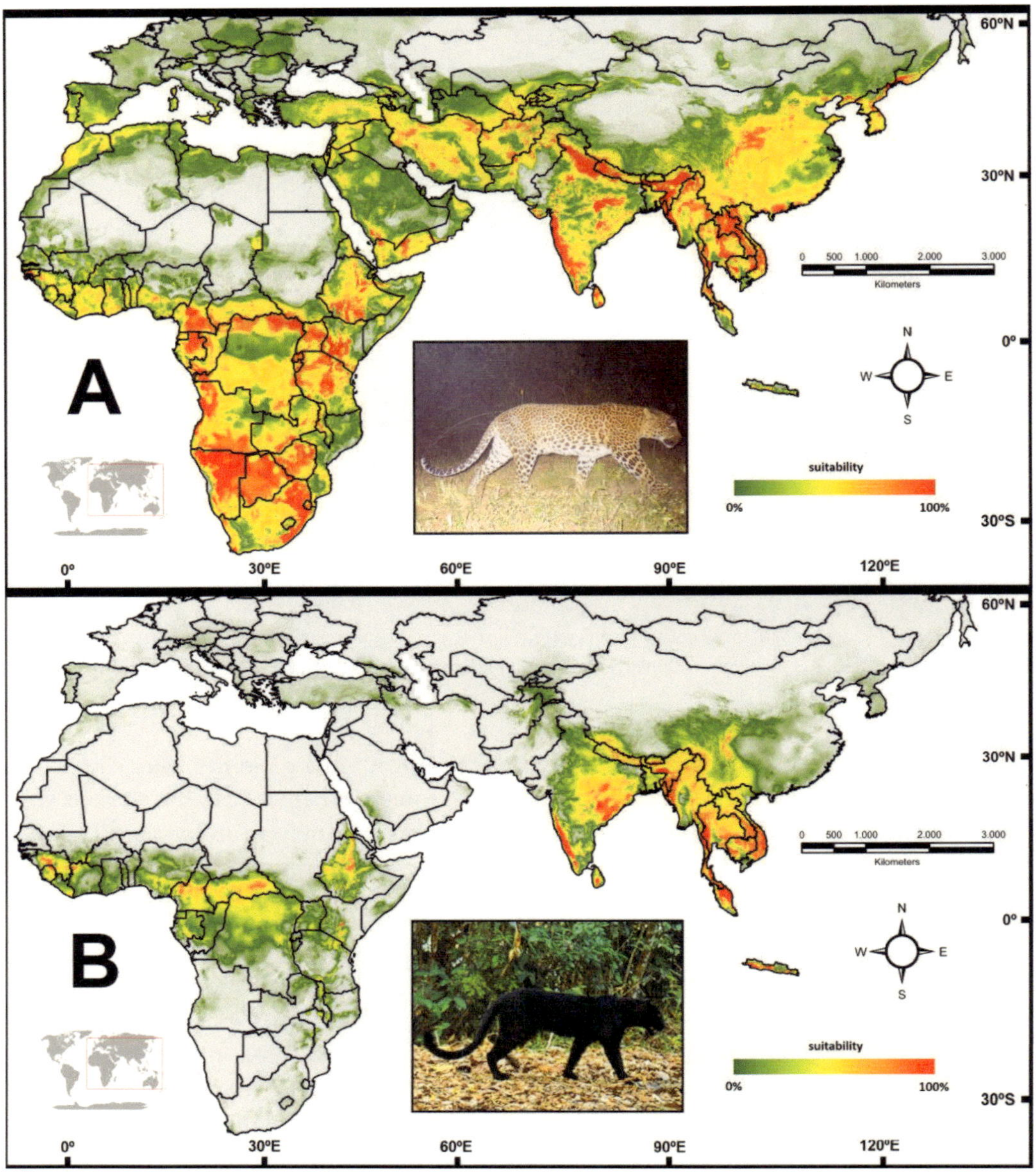

Figure 4. Potential distribution maps of the two coloration phenotypes analyzed in this study: (A) distribution of nonmelanistic leopards and (B) distribution of melanistic leopards. *Photos:* Andrew Stein and Bruce Kekule.

of Brazil were corroborated by this study, especially the Amazon and Cerrado areas (states of Amazonas, Pará, Mato Grosso, Goiás, and Minas Gerais; see Ref. [43]) and Caatinga (Serra da Capivara National Park [44, 45]). Additionally, new records were obtained for areas where there were no previous reports of melanistic animals (Colombia, Peru, Ecuador, and Costa Rica). When the presence of melanism across regions was assessed, we observed marked differences in its frequency among distinct biomes and ecoregions. For example, melanism can reach high frequencies in a single ecoregion, such as the Alto Paraná Atlantic Forest, Cerrado, and Caatinga (inside the Brazilian territory), whereas can present a complete lack in the Pantanal and Llanos (implying a significantly lower frequency of this phenotype in these regions).

Niche models generated here were considered satisfactory (AUC ≥ 0.9): nonmelanistic model (AUC training = 0.949, test = 0.941) and melanistic model (AUC training = 0.978, test = 0.955). The nonmelanistic model provided a good fit to know the present broad distribution model for jaguars in the Americas, indicating that it provides a suitable baseline against which to compare the melanism model (**Figure 5A**). The melanism model showed some similarities with the control model, but also some important differences, especially the low suitable habitats for occurrence in the Pantanal and Llanos ecoregions (**Figure 3B**). When were analyzed the environmental variables that were most influential on the models, we observed that predictors related to mean temperatures have the largest effect.

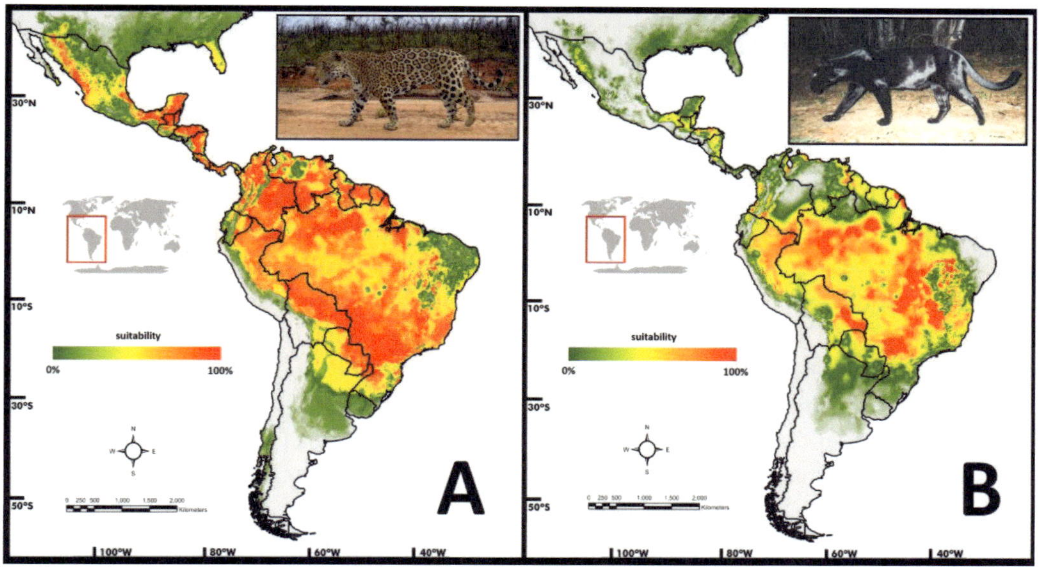

Figure 5. Potential distribution maps of the two coloration phenotypes analyzed in this study: (A) distribution of nonmelanistic jaguars and (B) distribution of melanistic jaguars. The species is considered almost as panmictic population [46] and near threatened by IUCN. *Photos:* Edsel Amorim Jr. and Leandro Silveira.

4. Discussion

The combination of statistical techniques associated with geoprocessing data has been used for some time in predictive models of ecology [47], especially in ecological niche models in the context of macroecological analysis [37, 41, 48, 49]. The models shown in **Figures 3** and **5** were designed to provide an analysis of the relative influence of environmental predictors on the geographic distribution of melanism in big cats. The main differences between nonmelanistic and melanistic models were restricted to predictors (moisture and temperature) related to thermoregulation in natural habitats, suggesting that the presence (or frequency) of melanism in panthers may be regulated by climatic features.

For a character to be recognized as adaptive, it must be derived and involved in the response to a selective agent [50], and in this context, it is interesting to determine if a polymorphism deviates from equilibrium expectations [51]. To elucidate biological issues related to melanism in natural populations and assess the relevance of different adaptive phenotypes, it is necessary to consider the relative importance of genetic drift and natural selection on the dynamics of different phenotypes in distinct landscapes [33, 52]. A selectively neutral phenotype should show a random pattern of variation among populations, while nonrandom patterns suggest the occurrence of selection (if populations are demographically connected). In the case of a stable polymorphism (such as melanism), an important issue to be considered is the phenotype frequency across different landscapes [53] because in some cases ecological variables describing a species range can predict genetic patterns [33]. That is the exactly pattern observed in leopards and jaguars, where the frequency is almost the same for both species in natural populations, and the geographic distribution is nonrandom in the two cases. This scenario supports the hypothesis of natural selection acting under the frequency and occurrence of melanism in these animals.

In fact, this study has provided a characterization of the spatial distribution of melanism in leopards and jaguars and demonstrated that such distribution is nonrandom. Recent analyses have indicated that melanism can reach very high frequencies in some leopard populations (e.g., Southeast Asia reported by [14]). In addition, there have been confirmed reports of melanistic leopards in India, Abyssinia, and Ethiopian Highlands, Java and Malaysia, Aberdare Mount Kenya, Highlands of Nepal, as well as a doubtful occurrence in South Africa [19, 26, 28, 54]. These observations are restricted to some areas and may support the hypothesis that melanism can provide an adaptive advantage in certain ecological conditions [5, 18]. At least four of the nine currently recognized leopard subspecies (based on [55]) are already cited in the literature as having confirmed records of melanistic animals: *P. p. pardus, P. p. fusca, P. p. delacouri*, and *P. p. melas*. However, the exact geographic range of this coloration phenotype has never been mapped in leopards in general, or in any of its subspecies.

Moist forests (especially in Southeast Asia) presented very high frequencies of melanistic leopards (e.g., 39 of 71 individuals [55%] in Southeast Asia), and more than 80% of the black animals in our database, a fivefold increase relative to the expected number based on the overall number of records. Furthermore, we found no confirmed melanistic leopards in the Middle East, Arabian Peninsula, Central China, and Russian Far East (**Figure 1**), nor any mention in the literature as to the presence of melanism in these regions, indicating that this polymorphism is really absent from these areas. Overall, there was a significant reduction in the frequency of melanism in some biomes that consist of open habitats or temperate forests. There thus is a clear pattern in which melanism tends to increase in tropical/subtropical moist forests and decrease in open/dry or temperate habitats. Such a result supports the classical hypothesis postulating an adaptive role for melanism, which would be favored in tropical and humid environments.

An alternative explanation is that variation in the melanism frequency could have been driven by demographic processes, including population structure and drift-induced differentiation. According to Ref. [55], significant geographic structure can be identified in leopards, indicating the existence of restricted historical gene flow among some portions of the range. This division formed the basis for currently recognized leopard subspecies, although in that study the authors noted that in some areas, such as African continent, Arabian Peninsula, Sri Lanka, and Java, the sampling was too sparse to identify clear-cut phylogeographic relationships. Nevertheless, the presence of demographic distinctiveness among at least the nine recognized species argues that such differentiation must be taken into account when comparing the frequency of melanism among regions. Although the possibility that demography has influenced the present-day frequency of melanism in leopards cannot be completely excluded, it is unlikely that it could explain most of the observed patterns, since each of the subspecies' ranges tends to contain a variety of habitats. Therefore, demographic effects caused by historical differentiation among subspecies would tend to obscure, rather than generate, the observed pattern of association between melanism and forested habitats.

Therefore, we consider that the most probable historical scenario for melanism in leopards is the emergence of the causative allele at a particular location and its dispersal throughout much of the species' distribution, suffering selection under the influence of varying environmental conditions in different landscapes, as well as genetic drift in some situations (e.g., founding of new populations during range expansion events). Since melanism in leopards is a recessive trait [12], it is plausible that its causative allele can disperse long distances over evolutionary time even across habitats where it could be selected against (e.g., deserts and grasslands). This is because the allele can remain "hidden" in heterozygosity for many generations when it is at low frequency, while at the same it could be lost in some areas due to an effect of genetic drift. Another possibility is that melanism arose in leopards more than once.

When the distribution of melanism in leopards is examined more closely, Southeast Asia emerges as a particularly interesting region. Our data support the findings reported by [14] and [15], showing that melanism is almost fixed in areas south of the Isthmus of Kra (Thailand/Malaysia). This study obtained only two records of nonmelanistic animals south of the Isthmus, while in more northerly areas both phenotypes appear at similar frequencies. This intriguing pattern may have been caused by some degree of demographic isolation across the Isthmus, which is consistent with the hypothesis that in the past (during the Last Glacial Maximum period, between 20,000 and 25,000 years ago, [56]) it operated as an effective ecological barrier restricting gene flow for several organisms. The fact that present-day landscapes appear to be similar on both sides of the Isthmus argue for a demographic, rather than selective explanation of the high frequency of melanism in the southern portion. However, the analysis of environmental predictors that influence the distribution of leopards in Southeast Asia revealed that moisture is the main factor that induces the presence of melanism, without correlations with landscape features (vegetation index). Given the observed support for an increased frequency of melanism in moist forests areas, it is interesting to discuss its potential causes in the light of classical hypotheses suggesting a selective advantage related to thermoregulation [57] in leopards.

The most part of melanistic jaguars was recorded in South America, especially in Brazil, Ecuador, and Peru territories. The species potential distribution map and melanism distribution map indicate a high habitat suitability in moist areas. As we know, the Amazon region has a large size and high population density of jaguars and can be considered a core habitat for the species [58]. In addition, the Brazilian Atlantic Forest region, despite the recent population decline [59–61] and the local loss of genetic diversity in jaguars [62], still has remaining populations. The presence of melanism in Brazilian moist and dry forests and Cerrado areas had already been documented and can reach high local frequencies in some remnant areas [44, 63]. In contrast, flooded grasslands and savanas such as Pantanal (Brazil) and Llanos (Colombia/Venezuela) present no records of melanistic jaguars, indicating that this polymorphism is really absent in these areas.

Although jaguars are often documented in North and Central Americas [19], there were historically only four records of melanistic individuals in these regions. The only record of melanism from North America prior to this study was a black female photographed in 2004, in the El Fuerte River Valley, near Sinaloa, Mexico [64]. In Central America, there are two records of melanistic animals from Belize: Ek Balan and El Rancho Grande River (previously reported as possible by [43]), and one record from Costa Rica [65]. Remaining populations of the species have been recently identified in the northern portion of its distribution, in the southern United States [66, 67], but there has been no record of melanism in these areas. The generated models showed high suitability for melanism occurrence in Belize and Costa Rica and low suitability in Mexico.

Previous studies have shown that jaguars possess low levels of geographic structure on a wide range of scale [46]. Phylogeographic analyses indicated that there were no impassable historical barriers to gene flow throughout the species' range. Only a few historical barriers to dispersal were inferred at this scale, such as the Amazon River, whose influence was much stronger on the female-inherited mitochondrial DNA than on nuclear markers. These results suggested that the species has behaved historically almost as a panmictic population, which argues against the possibility that founder effects and/or high genetic drift at a regional scale could have induced the observed nonrandom patterns in the distribution of melanism. Moreover, there is so far no evidence of historical bottleneck events [46], which could have exacerbated genetic drift and thus lead to large scale increases in the frequency of melanism. In this context, it may be noted that melanism is present with high suitability on both sides of the Amazon River, suggesting that it has not been affected by the historical (albeit incomplete) restriction to gene flow inferred with molecular markers. Given these considerations, this study conclude that the most probable scenario is the jaguar melanism allele arose at a particular location and dispersed throughout the species' distribution, with its regional frequency being influenced at least partially by natural selection related to environmental predictors that vary across different landscapes. In conclusion, the geographic variation observed here provides evidence for the existence of natural selection in this system and provides some hints as to its nature.

5. Conclusions

Given the current genetic evidence indicating that population structures in leopards and jaguars have historically comprised a broadly connected population across its distributions, it would be expected to find a random distribution pattern of melanism throughout their ranges if this variant was selectively neutral. The results of this study indicate that this is not the case and demonstrate that its distributional pattern is nonrandom. Although the observed punctual cases of increased frequency of melanism may be indeed derived from recent drift-induced shifts in locally isolated populations, we do not find support for a drift-based explanation for the observed pattern, and thus favor a scenario implying an effect of natural selection under polymorphic phenotypes in big cats such as melanism. Moreover, the suitability maps generated here showed differences in the distribution of melanistic records relative to the ancestral phenotype (spotted) and suggested that at least some of the underlying differences could be related to environmental predictors, possibly to thermoregulation issues. Overall, this study contributed to address a question that has circulated anecdotally for almost 200 years in the scientific literature but remained blurred in zoology, particularly in wild felids. The results open up new avenues for investigating this polymorphism in other mammal species. In this context, it is relevant to note that the ecology of some species can drive genetic diversity, and phenotypes in natural populations can present spatial patterns of variation associated with adaptation to some environments, regulated by natural selection. By connecting coloration diversity to environmental information which may have a direct or indirect influence on the adaptive significance of coat color variants, it is possible to investigate evolutionary processes involved in the generation and maintenance of polymorphic phenotypes in nature.

Acknowledgements

Lucas G. da Silva would like to thank Eduardo Eizirik, Ronaldo Morato, Laury Cullen Jr., Benoit DeThoisy, Emiliano Ramalho, Esteban Payán, Leandro Silveira, Edsel Amorim Jr., Rony Garcia-Anléu, José Moreira Ramirez, Kae Kawanishi, Agustin Paviolo, Philipp Henschel, Andrew Kittle, Arezoo Sanei, Dale Miquelle, Andrew Stein, Anjali Watson, Bruce Kekule, Ricardo Machado, Katia Barros Ferraz, Francesca Palmeira, Taiana Haag, Daniel Rocha, Tiago Freitas, Rogério Cunha de Paula, Alexandre Uezu, Micheline Vergara, Kristofer Helgen, Eileen Westwig, Esther Langan, Renata Bornholdt, Linda Gordon, Danis Sana, Analice Calaça, Samuel Perez, Alexandre Vogliotti, Marina Silva, Flavia Conde, Leonardo Viana, Carlos Benhur Kasper, Beatriz Beisiegel, João Feliz Moraes, Thaiane Weinert, Milind Pariwakam, Biswajit Mohanty, Kashmira Kakati, Anhar Harahap, Vidya Atreya, Jane Budd, Arash Ghoddousi, Ezekiel Fabiano, Peter Crawshaw Jr., Alexander Reebin, Guy Balme, David Stanton, Yury Shibnev, Dipankar Ghose, João Feliz Moraes, Martín Alejandro Montés, and Mario DiBitteti for location records and/or scientific support; to American Museum of Natural History, Smithsonian Institution, Jaguar Conservation Fund, Instituto Pro-Carnivoros, CENAP ICMBio Brazil, Instituto Mamirauá, Panthera, Wildlife Conservation Society, World Wild Fund, Asian Leopard Specialist Group, Cape Leopard Trust, IUCN Cat Specialist Group, Instituto Biotrópicos, and PUCRS for scientific support; and to CNPq/Brazil and FACEPE for financial support.

Author details

Lucas Gonçalves da Silva[1,2*]

*Address all correspondence to: lucas_gonc@yahoo.com.br

1 Federal Rural University of Pernambuco, UFRPE, Brazil

2 Pontifical Catholic University of Rio Grande do Sul, PUCRS, Brazil

References

[1] Gloger CWL. Das abändern der vögel durch einfluss des climas [The evolution of birds through the impact of climate]. Breslau: August Schulz; 1833

[2] Poulton EB. The Colors of Animals, their Meaning and Use, Especially Considered in the Case of Insects. London: Kegan Paul, Trench & Trübner; 1890

[3] Beddard FE. Animal Coloration: Color and Markings of Animals. 2nd ed. London: Swan Sonnenschein; 1985

[4] Cott HB. Adaptive Coloration in Mammals. London: Methuen, United Kingdom; 1940

[5] Majerus MEN. Melanism - Evolution in Action. Oxford University Press, United Kingdom; 1998

[6] Caro T. The adaptive significance of coloration in mammals. BioScience. 2005;**55**(2):125-136

[7] Eizirik E, O'Brien SJ. Evolution of the melanism in the Felidae. Cat News. 2003;**38**:37-39

[8] Nachman MW, Hoekstra HE, D'Agostino SL. The genetic basis of adaptive melanism in pocket mice. Proceedings of the National Academy of Sciences USA. 2003;**100**:5268-5273

[9] Hoekstra HE. Genetics, development and evolution of adaptive pigmentation in vertebrates. Heredity. 2006;**97**:222-234

[10] Ortolani A, Caro T. The adaptive significance of color patterns in carnivores. In: Gittleman JL, editor. Carnivores Behavior, Ecology and Evolution. Vol. 2. Cornell University Press; 1996

[11] Eizirik E, Yuhki N, Johnson WE, Raymond M, Hannah SS, O'Brien SJ. Molecular genetics and evolution of melanism in the cat family. Current Biology. 2003;**13**:448-453

[12] Schneider A, David VA, Johnson WE, O'Brien SJ, Barsh GS, Maymond MM, Eizirik E. How the leopard hides its spots: *ASIP* mutations and melanism in wild cats. Plos One. 2012;**7**:e50386

[13] Schneider A, Henegar C, Day K, Absher D, Napolitano C, Silveira L, David VA, O'Brien SJ, Raymond MM, Barsh GS, Eizirik E. Recurrent evolution of melanism in South American felids. Plos Genetics. 2015;**10**:e1004892

[14] Kawanishi K, Sunquist ME, Eizirik E, Lynam AJ, Ngoprasert D, Wan Shahruddin WN, Rayan DM, Sharma DSK, Steinmetz R. Near fixation of melanism in leopards of the Malay Peninsula. Journal of Zoology. 2010;**282**(3):201-206

[15] Hedges L, Lam WY, Arceiz AC, Rayan DM, Laurance WF, Latham CJ, Saaban S, Clements GR. Melanistic leopards reveal their spots: Infrared camera traps provide a population density estimate of leopards in Malaysia. The Journal of Wildlife Management. 2015;**79**:846-853

[16] Allen WL, Cuthill IC, Samuel NE, Baddeley R. Why the leopard got its spots: Relating pattern development to ecology in felids. Proceedings of the Royal Society: Biological Sciences. 2010;**277**, online

[17] Ulmer FA. Melanism in the Felidae, with special reference to the genus *Lynx*. Journal of Mammalogy. 1941;**11**:185-188

[18] Majerus MEN, Mundi NI. Mammalian melanism: Natural selection in black and white. Trends in Genetics. 2003;**19**(11):585-588

[19] Sunquist M, Sunquist F. Wild Cats of the World. University of Chicago Press, United States; 2002

[20] Jacobson AP, Gemgross P, Lemeris JR, Schoonover RF, Anco C, Breitenmoser C, Durant SM, Farhadinia MS, Henschel P, Kamler JF, Laguardia A, Garcia S, Stein AB, Dollar L. Leopard (*Panthera pardus*) status, distribution, and the research efforts across its range. PeerJ. 2016;**4**:e1974

[21] Sanderson EW, Redford KH, Chetkiewicz CB, Medellin RA, Rabinowitz A, Robinson JG, Taber AB. Planning to save a species: The jaguar as a model. Conservation Biology. 2002;**16**(1):58-72

[22] Haag T, Santos AS, Valdez FP, Sana D, Silveira L, Cullen Jr. L, De Angelo C, Morato RG, Crawshaw Jr. PG, Salzano F.M, Eizirik E. Molecular tracking of jaguar melanism using faecal DNA. Conservation Genetics. 2010;**11**(3):1239-1242

[23] Nelson EW, Goldman EA. Revision of the jaguars. Journal of Mammalogy. 1933;**14**(3): 221-240

[24] Robinson R. Homologous genetic variation in the Felidae. Genetica. 1976;**46**:1-31

[25] Dittrich L. Die verenbung des melanismus beim jaguar (*Panthera onca*). Zoologische Garten. 1979;**49**:417-428

[26] Silva LG. Análise da distribuição espacial do melanismo na família Felidae em função de condicionantes ambientais. [PhD dissertation]. Pontifícia Universidade Católica do Rio Grande do Sul; 2014

[27] Silva LG, Oliveira TG, Kasper CB, Cherém JJ, Moraes Jr. EA, Paviolo A, Eizirik E. Biogeography of polymorphic phenotypes: Mapping and ecological modelling of coat colour variants in an elusive Neotropical cat, the jaguarundi (*Puma yagouaroundi*). Journal of Zoology. 2016;**299**:295-303

[28] Silva LG, Kawanishi L, Henschel P, Kittle A, Sanei A, Reebin A, Miquelle D, Stein AB, Watson A, Kekule LB, Machado RB, Eizirik E. Mapping black panthers: Macroecological modeling of melanism in leopards (*Panthera pardus*). Plos One. 2017;**12**(4): e0170378

[29] Silva LG, Payán E, Crawshaw Jr. PG, Silveira L, Cullen Jr. L, Morato RG, Moraes Jr. EA, Ramirez JM, DeThoisy B, Anléu RG, Ramalho EE, Sana D, Eizirik E. A biogeographic assessment of jaguar (Panthera onca) melanism: Ecological modeling reveals evidences of adaptive differences in distinct biomes. In review. 2017

[30] Olson DM, Dinerstein E, Wikramanayake ED, Burgess ND, Powell GV, Underwood EC, D'Amico JA, Itoua I, Strand HH, Morrison JC, Loucks CJ, Allnutt TF, Ricketts T, Kura Y, Lamoreux JF, Wettengel W, Hedao P, Kassem KR. Terrestrial ecoregions of the world: A new map of life on earth. BioScience. 2001;**51**:933-938

[31] Kumar S, Stohlgren TJ. Maxent modeling for predicting suitable habitat for threatened and endangered tree *Canacomyrica monticola* in New Caledonia. Journal of Ecology and Natural Environment. 2005;**1**:94-98

[32] Raes N, Roos MC, Slik JWF, Van Loon EE, Steege H. Botanical richness and endemicity patterns of Borneo derived from species distribution models. Ecography. 2009;**32**:180-192

[33] Mukherjee S, Krishnan A, Tamma K, Home C, Navya R, Joseph S, Das A, Ramakrishnan U. Ecology driving genetic variation: A comparative phylogeography of jungle cat (*Felis chaus*) and leopard cat (*Prionailurus bengalensis*) in India. PlosOne. 2010;**5**(10):e13724

[34] Philips SJ, Anderson RP, Schapired RE. Maximum entropy modeling of species geographic distributions. Ecological Modeling. 2006;**190**:231-259

[35] Waltari E, Hijmans RJ, Peterson AT, Nyari AS, Perkins SL, Guralnick RP. Locating Pleistocene refugia: Comparing phylogeographic and ecological niche model predictions. PlosOne. 2007;**7**:e563

[36] Philips SJ, Dudik M. Modeling of species distributions with Maxent: New extensions and a comprehensive evaluation. Ecography. 2008;**31**:161-175

[37] Elith J, Phillips SJ, Hastie T, Dudik M, Chee YE, Yates CJ. A statistical explanation of MaxEnt for ecologists. Diversity and Distributions. 2010;**17**:43-57

[38] Merow C, Smith MJ, Silander JA. A practical guide to Maxent for modeling species' distributions: What it does, and why inputs and settings matter. Ecography. 2013;**36**:1058-1069

[39] Pearson RG. Species' Distribution Modeling for Conservation Educators and Practitioners. American Museum of Natural History Press, United States; 2007

[40] Ferraz KMPMB, Ferraz SFB, Paula RC, Beisiegel B, Breitenmoser C. Species distribution modeling for conservation purposes. Natureza & Conservação. 2012;**10**(2):214-220

[41] Calabrese JM, Certain G, Kraan C, Dormann CF. Stacking species distribution models and adjusting bias by linking them to macroecological models. Global Ecology and Biogeography. 2013; **23**:99-112.

[42] Plackett RL. Karl Pearson and the chi-squared test. International Statistical Review. 1983;**51**(1):59-72

[43] Meyer JR. Black Jaguars in Belize? A Survey of Melanism in the Jaguar (*Panthera onca*) [Internet]. 1994. Available from: http://biological-diversity.info/Black_Jaguar.htm (Accessed: 01-September-2013)

[44] Perez SEA. Ecologia da onça-pintada nos parques nacionais Serra da Capivara e Serra das Confusões, Piauí. [PhD dissertation] Brasília University; 2008

[45] Silveira L, Jácomo ATA, Astete S, Sollmann R, Tôrres NM, Furtado MM, Marinho-Filho J. Density of the near threatened jaguar *Panthera onca* in the Caatinga of north-eastern Brazil. Fauna & Flora International Oryx. 2009;**44**:104-109

[46] Eizirik E, Kim JH, Raymond MM, Crawshaw Jr. PG, O'Brien SJ, Johnson WE. Phylogeography, population history and conservation genetics of jaguars (*Panthera onca*, Mammalia, Felidae). Molecular Ecology. 2001;**10**:65-79

[47] Guisan A, Zimmermann NE. Predictive habitat distribution models in ecology. Ecological Modelling. 2000;**135**:147-186

[48] Carnaval AC, Moritz C. Historical climate modelling predicts patterns of current biodiversity in the Brazilian Atlantic forest. Journal of Biogeography. 2008;**35**:1187-1201

[49] Loiselle BA, Graham CH, Goerck JM, Ribeiro MC. Assessing the impact of deforestation and climate change on the range size and environmental niche of birds species in the Atlantic forests, Brazil. Journal of Biogeography. 2010;**37**:1288-1301

[50] Futuyma DJ. Evolution. 2nd ed. Sinauer Associates Inc., United States; 2009

[51] Kreitman M. Methods to detect selection in population with application to the human. Annual Reviews of Genomics and Human Genetics. 2000;**01**:539-559

[52] Lande R. Natural selection and random genetic drift in phenotypic evolution. Evolution. 1976;**30**:314-334

[53] Novembre J, DiRienzo A. Spatial patterns of variation due to natural selection in humans. Nature Genetics. 2009;**10**:745-755

[54] Thapa K, Pradhan NB, Barker J, Dhakal M, Bhandari AR, Gurung GS, Rai DP, Thapa GJ, Shrestha S, Singh GR. High elevation records of a leopard cat in the Kangchenjunga conservation area, Nepal. Catnews. 2013;**58**:26-27

[55] Uphyrkina O, Johnson WE, Quigley H, Miquelle D, Marker L, Bush M, O'Brien SJ. Phylogenetics, genome diversity and origin of modern leopard, *Panthera pardus*. Molecular Ecology. 2001;**10**:2617-2633

[56] Clark PU, Dyke AS, Shakun JD, Carlson AE, Clark J, Wohlfarth B, Mitrovica JX, Hostetler SW, McCabe AM. The last glacial maximum. Science. 2009;**325**:710-714

[57] Burtt EH, Ichida JM. Gloger's rule, feather-degrading bacteria, and color variation among song sparrows. The Condor. 2004;**106**:681-686

[58] Tobler MW, Percastegui SEC, Hartley AZ, Powell GVN. High jaguar densities and large population sizes in the core habitat of the southwestern Amazon. Biological Conservation. 2013;**159**:375-381

[59] Paviolo A, DeAngelo C, DiBlanco Y, DiBitteti M. Jaguar *Panthera onca* population decline in the upper Paraná Atlantic forest of Argentina and Brazil. Fauna & Flora International Oryx. 2008;**42**(4):554-561

[60] Galetti M, Eizirik E, Beisiegel B, Ferraz KMPMB,Cavalcanti S, Araujo AC, Crawshaw Jr. PG, Paviolo A, Galetti Jr. PM, Jorge ML, Marinho-Filho J, Vercillo U, Morato RG. Atlantic rainforest's jaguars in decline. Science. 2013;**342**:930

[61] Paviolo A, DeAngelo C, Ferraz KMPMB, Morato RG, Pardo JM, Araujo ACS, Beisiegel B, Lima F, Sana D, Silva MX, Velazquez MC, Cullen Jr. L, Crawshaw Jr. PG, Jorge ML, Galetti Jr. PM, DiBitetti M, Paula RC, Eizirik E, Aide TM, Cruz P, Perilli ML, Souza ASMC, Quiroga V, Nakano E, Pinto FR, Fernandez S, Costa S, Moraes Jr. EA, Azevedo F. A biodiversity hotspot losing its top predator: The challenge of jaguar conservation in the Atlantic forest of South America. Scientific Reports. 2016;**6**. doi: srep37147

[62] Haag T, Santos AS, Sana D, Morato RG, Cullen Jr. L, Crawshaw Jr. PG, DeAngelo C, DiBitteti M, Salzano FM, Eizirik E. The effect of habitat fragmentation on the genetic structure of a top predator: Loss of diversity and high differentiation among remnant populations of Atlantic Forest jaguars (*Panthera onca*). Molecular Ecology. 2010;**19**(22):4906-4921

[63] Cullen Jr. L, Abreu KC, Sana D, Nava AF. As onças-pintadas como detetives da paisagem no corredor do Alto Paraná, Brasil. Natureza & Conservação. 2006;**3**(1):43-58

[64] Dinets V, Polechla Jr. P. First documentation of melanism in a jaguar from northern Mexico. Cat News. 2007;**42**:17

[65] Bolaños CS, Montalvo V, Fuller TK, Carrillo E. Records of black jaguars at Parque Nacional Barbilla, Costa Rica. Cat News. 2015;**62**:38-39

[66] Grigione M, Scoville A, Scoville G, Crooks K. Neotropical cats in southeast Arizona and surrouding areas: Past and present status of jaguars, ocelots and jaguarundis. Mastozoología Neotropical. 2007;**14**(2):189-199

[67] McCain EB, Childs JL. Evidence of resident jaguars (*Panthera onca*) in the southwestern United States and the implications for conservation. Journal of Mammalogy. 2008;**89**(1):1-10

Chapter 7

The Last Coastal Jaguars of Ecuador: Ecology, Conservation and Management Implications

Miguel Saavedra Mendoza, Paúl Cun,
Eric Horstman, Sonia Carabajo and Juan José Alava

Additional information is available at the end of the chapter

http://dx.doi.org/10.5772/intechopen.69859

Abstract

Ecuador is one of the top countries with the highest biodiversity indexes in the planet. Among the mammal species inhabiting tropical forests along Ecuador's coast, wild cats such as ocelots (*Leopardus pardalis*), jaguarundis (*Puma yagouaroundi*), cougars (*Puma concolor*) and jaguars (*Panthera onca*) are a key group of carnivores deserving critical consideration because these species are facing several anthropogenic threats and conservation challenges. Of particularly attention is the critically endangered subspecies of jaguar (*Panthera onca centralis*) from the Ecuadorian coast. Despite this species is the largest cat in Ecuador's coastal tropical forests and demanding large territories to survive, little is known about its population and conservation status. In most forests along Ecuador's coast, habitat loss due to deforestation and fragmentation, poaching of prey and illegal hunting threaten the survival of jaguars and questions linger about its ecology and population health. Based on recent field observations using transects and deployment of camera traps, as well as surveys conducted with the local community in and around Cerro Blanco Protected Forest and surrounding areas of the Cordillera Chongón-Colonche Mountain Range, we advance the state of the ecological knowledge of coastal jaguar populations with conservation implications of its threatened habitat and long-term survival in Ecuador.

Keywords: jaguar, *Panthera onca centralis*, population ecology, tropical forests, anthropogenic threats, conservation, Ecuador, South America

1. Introduction

The jaguar (*Panthera onca*) is the largest wild cat of the Americas and the supreme predator of the Amazon River Basin, the largest flooded tropical rain forest in South America. The jaguar has a broad distribution in the Americas, from the southwest of Unites States and Mexico, through Central America, to the north of Argentina [1–5]. Because the rates of deforestation are high in Latin America and associated habitat fragmentation have isolated jaguar populations, they are more susceptible to human persecution as well as human competition for jaguar's prey [2, 5]. While the known geographical distribution range and are of occupancy originally occupied by jaguars has currently been reduced by 48–55% in the Americas region in the last century [6], it has been estimated that about 27% of the jaguar range exhibits depleted major wild prey [7].

In Ecuador, this species inhabits the lowlands and foothills of tropical forest and fragmented areas from tropical (i.e., 800–1000 m above mean sea level, AMSL) and subtropical (i.e., from 800 to 1000 m to 1800–2000 m AMSL) regions located to the western and eastern the Andean chain range, respectively [3, 8–13]. Of particularly attention is the critically endangered subspecies of jaguar (*Panthera onca centralis*) from western Ecuador inhabiting humid and dry tropical forests still persisting along the coast, mainly in protected areas such as the Cotacachi-Cayapas Ecological Reserve (Esmeraldas province), Machalilla National Park (Manabí province) and Cerro Blanco Protected Forest in Guayas province [14–18]. While the coastal jaguar has been considered to be extirpated from Machalilla National Park [18], where the available habitat for jaguar has been reduced by 80% due to deforestation [8] with the last records being reported about 20 years ago [19], the species stills exist at the north part of the Cotacachi-Cayapas Ecological Reserve [17, 20], as well as in BPCB [14, 16].

The presence of jaguars has recently been confirmed in Cerro Blanco Protected Forest located at the southeast of the "Cordillera" Chongón Colonche Range (i.e. Chongón Colonche Protected Forest) on the central coast of Ecuador [14, 16]. Likewise, the plausible existence of jaguars has been suggested in areas with small woodland remnants around the La Ercilia village, Los Ríos province, on the coast of Ecuador (M. Saavedra-Mendoza, *personal communication*, 2017). Other wild cat species inhabiting forests along Ecuador's coast include the ocelot (*Leopardus pardalis*), oncilla (*Leopardus tigrinus*), margay (*Leopardus wiedii)*, jaguarundi (*Puma yagouaroundi*) and puma (*Puma concolor*) [11, 12].

As an emblematic species, the jaguar is a fundamental icon of the ancient cultures and traditional knowledge of aboriginal communities and farmers in Ecuador. The archaeological findings echo that the first records of the jaguar's presence in Ecuador dated back to the artistic manifestations left by the pre-Columbian cultures [8, 13]. The presence of ceramics dating more than 3000 years old clearly represent jaguars, occupying an important place in the cultures of the pre-Columbian Ecuador, where it was worshiped (e.g., "*otorongo*" in the Antisuyo period: 530–468 B.P.) and represented a symbol of power and connection between men, mother nature and gods [21, 22]. At present, several indigenous tribes (e.g., the Waorani, The Cofán, The Awá) in Ecuador maintain myths and related beliefs with the jaguar [13]. The jaguar was named "El tigre" by Spanish explorers, who recalled the tiger in Asia [1]. Actually, the jaguar has also been identified with "the tiger" in coastal Ecuador by farmers from rural communities for generations due to its alleged similar appearance and ferocity.

While Ecuador is one of the top megadiverse countries with one of the highest biodiversity indexes in the planet [23], deforestation and several other anthropogenic threats such as habitat loss and fragmentation due to urbanization and agriculture, illegal poaching and hunting, mining, oil exploitation, and pollution jeopardize the survival of several tropical mammals [24]. Particularly, the jaguars dwelling in the last remnants of tropical forests along Ecuador's coast are threatened due to several human made threats [15, 16, 18]. Contrasting to the substantial field research invested and dedicated to the subspecies of jaguar from the Amazon basin jungle (*P. onca onca*), little is known on the population and conservation status of coastal jaguars from Ecuador. Despite this species being the largest cat in Ecuador's coastal tropical forests and demanding large territories to survive, questions linger on whether a healthy population of jaguars still remains along the coastal region. In general, Ecuador's wild felines are a key group of carnivores deserving critical consideration because they are facing several environmental stressors and conservation challenges. In this chapter, we review the recent conservation status of coastal jaguars and contribute with new ecological data using as study case the population of the last jaguars inhabiting Cerro Blanco Protected Forest in Ecuador.

2. Materials and methods

2.1. Study area

The study area comprises the tropical forests of Cerro Blanco Protected Forest (BPCB hereafter) and the surrounding areas of the "Cordillera" Chongon-Colonche Range in Guayas province (**Figure 1**). The BPCB total area is 6000 ha and is located at the southern part (last extension) of the "Cordillera" Chongon-Colonche Range, close to Guayaquil City. Based on the floristic inventory the National Herbarium of Ecuador, BPCB possesses five categories of potential natural vegetation: Dry Plain Forest, Dry Rocky Forest, Humid Ravines Forest, Subhumid Plateau Forest, and Subhumid Montane Forest [25, 26]. The Chongon-Colonche Range is a mountain range on the central pacific coast of Ecuador, located in the provinces of Guayas, Santa Elena and Manabí. The Cordillera" Chongon-Colonche Range harbours the Chongón Colonche Protected Forest (≈86,000 ha), which is biogeographically found within the Tumbes-Choco-Magdalena biodiversity hotspot [23, 27] and part of the Chongón Colonche-Machalilla biological corridor, being one of the last remnants of native flora and vegetation in the coast. The forests still remaining in this mountain range contain high levels of biodiversity and was given the category of an Important Bird Area (IBA) [27, 28]. The Chongón Colonche Range is of paramount importance because of the ecological functions and environmental wellbeing and services that it provides to wildlife and local communities. This mountain range harbours high biodiversity and interconnects potential biological corridors for threatened species with wide distribution ranges. In fact, the Chongón-Colonche Range is one of the few coastal regions of Ecuador that still offers potential habitat suitability to big cats, among them the jaguar of the coast, *P. onca centralis* [16].

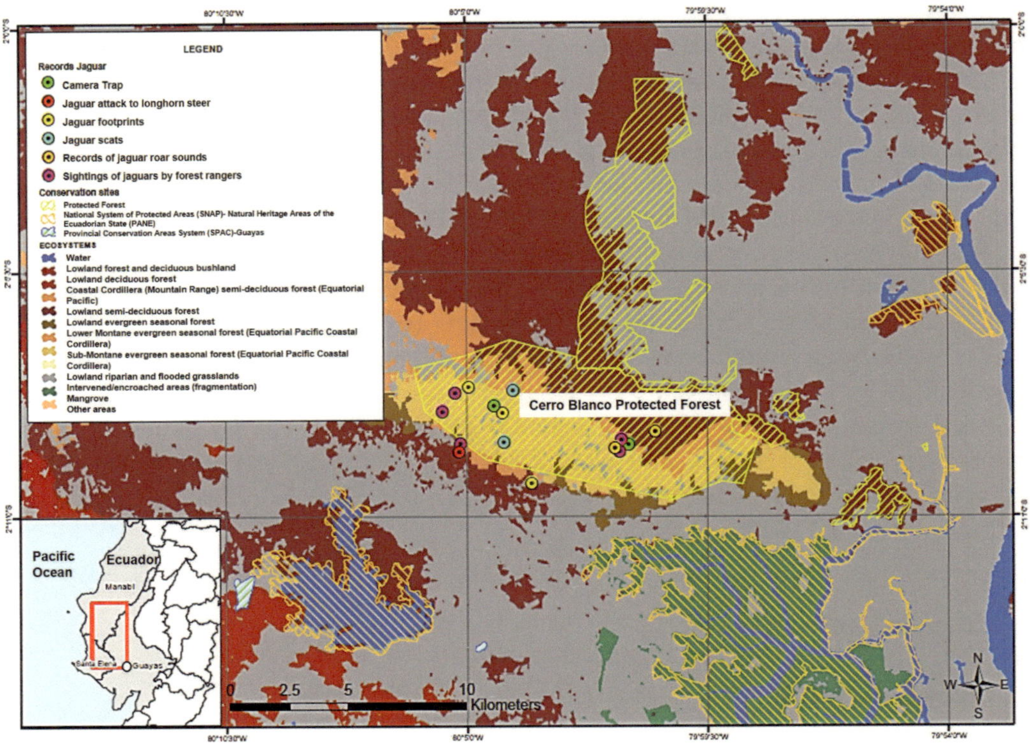

Figure 1. Map illustrating the study area and field observations and records of coastal jaguars (*P. onca centralis*) reported during the sampling and field work along transects as well as rangers encounters in and around the BPCB (area encircled with a white circle in inset map), Guayas Province, Ecuador. The inset map (bottom left corner) displays the Ecuadorian coast (grey area) with the approximate boundaries (red rectangle) of the Cordillera Chongón Colonche Mountain Range (Guayas, Santa Elena and Manabí provinces) and part of Machalilla National Park (Manabí Province).

2.2. Field sampling and observations

Field sampling and observations were focused on monitoring indirect records following the methods described by Cuesta et al. [29]. Thus, from June 22, 2008, to February 9, 2009, we deployed sampling efforts searching for excrements, footprints, feeding and resting sites. Potential direct observations were considered in sites, where the animals have left their tracks (e.g., footprints), following the methodology by Aranda [30]. Transects and camera trap were deployed in BPCB and surrounding areas of the Chongon-Colonche Mountain Range [16].

2.3. Experimental design and spatial sampling for transects

Five study sites were selected to deploy transects in the BPCB (**Figure 1**), described as follows: (a) transect 1 is characterized by being at a height of about 300 m above mean sea level (AMSL); at this level, it is common to find mist during most of the year with the presence of occasional "garuas" in the dry season (i.e. as reference, the "caseta Pigio" (Pigio guard station) is located very close to the study site); (b) transect 2 is part of a ravine, within which most of its extension is included and surrounded by several other small ravines along its route, converging in it; this sites usually maintains water during the dry season (i.e. as a reference, its initial point (Point 0) is very close to the "caseta three Bocas" (3 Bocas guard station)); (c) transect 3 extends along a path leading from the installations of the Holcim quarry facilities (i.e., authorized Swiss quarry company to exploit limestone) adjacent to the BPCB towards the "caseta Papagayo" (Papagayo guard station), at 150 m AMSL, and it is a dirt road that is rarely used with several ravines perpendicularly converging along its track; (d) transect 4 extends along a trail that connects the "Papagayo hut" to the "Jaguar Caseta" (Jaguar guard station), starting with a steep slope of about 200 m, followed by a flat trail along all its path; transect 4 altitude is at 300 m AMSL and ends up at a ravine, called "Quebrada Cóndor," harbouring important remnants of water sources in several sectors; and, (e) transect 5 is a dirt road between the Jaguar and Pigio stations, at about 300 m AMSL. **Table 1** shows the geographical coordinates for transected deployed in BPCB.

The monitoring was conducted along each transect (2 km length) and were marked every 50 m with orange fluorescent tape. The criteria to select transects were mainly based on trails that were rarely used by humans, as well as sectors contiguous to ravines and paths made by native fauna. Furthermore, all sites were selected relying on field information and traditional knowledge from rangers and ancestral users assuring that they have observed feline records in these sites from the forest. The field observations and data collection were conducted during diurnal hours with a total sampling effort of 3 days per transect. In addition, field inspections were carried out at the installations of the Holcim quarry company, located at 2.17°S and 80.06°W, following the invitation of staff members of the environmental department who observed footprints in sites around the quarry. They also provided evidence of past sightings and testimonies of witnesses that complemented the information of presence of jaguars in BPCB.

Transects/stations	**Point 0**		**Point 1000**		**Point 2000**	
	Longitude	**Latitude**	**Longitude**	**Latitude**	**Longitude**	**Latitude**
Transect 1 Reference: trail to Pigio station	80.03°W	2.15°S	80.04°W	2.15°S	80.045°W	2.15°S
Transect 2 Reference: Ravine to 3 Bocas station	80.01°W	2.15°S	80.01°W	2.16°S	80.015°W	2.16°S
Transect 3 Reference: Trail to Papagayo station-Holcim	80.07°W	2.17°S	80.08°W	2.17°S	80.08°W	2.16°S
Transect 4 Reference: Trail to Papagayo station-Jaguar station	80.08°W	2.14°S	80.09°W	2.15°S	80.09°W	2.155°S
Transect 5 Reference: Trail to Jaguar station-Pigío station	80.07°W	2.14°S	80.07°W	2.14°S	80.06°W	2.15°S

Table 1. Geographical coordinates and spatial data of transects used in the Cerro Blanco Protected Forest (BPCB).

2.4. Camera trap deployment

Cuddeback digital camera traps were deployed in several sectors along transects at BPCP. This digital device has 3.1 megapixels with daylight and 1.3 megapixels at night (in this period the photographs were in black and white), staying active 24 hours a day. Specifically, a camera trap was installed at a site on a probable jaguar route in the vicinity of a guard stations named "caseta Cusumbo." The photographic records obtained from larger cats allowed us to identify species and determine the population estimate for jaguar in BPCB by capturing and recapturing individuals through the identification of natural jaguar marks with the methodology proposed elsewhere [31, 32]. Similarly, the photo-records of medium and large mammals were used to determine the tropical fauna and/or potential prey of big cats, including jaguars.

2.5. Surveys with the local community and forest rangers

Interviews, anecdotic communications and traditional knowledge from forest rangers, fishers and residents of BPBP and "Cordillera" Chongón Colonche Range, as well as coastal communities inhabiting villages and towns around the study areas were taking into consideration for this work. This qualitative information was collated from the early 1990s to April 2017.

2.6. Preliminary assessment anthropogenic threats

With the aim to provide a snapshot assessment of the status information on human made impacts affecting the conservation and survival of coastal jaguar from BPCB and "Cordillera" Chongón-Colonche Mountain Range, we compiled the best information available for anthropogenic threats [8, 13, 15, 16], and then we developed a rating scheme based on the authors' knowledge as well as best technical judgment, as shown in **Table 2**. The ratings indicate as

Rating*	Meaning
	Status, trend, data, and/or actions provide contradictory or inconclusive information. Actions are needed to move into positive status and trend and avoid negative status and trend.
	1. The status of the species was or is healthy according to available data or/and best judgment 2. The population trend is positive if known 3. Some data are available; and/or 4. Actions to address or mitigate impacts are or will be underway and are likely to be potentially effective. Actions should be taken to maintain positive status and/or trend.
	1. Negative Impacts and threats were or are high risk and have resulted in a threatened or endangered status of the species 2. Improvements are uncertain, minor, or slow; and/or 3. Actions to address or mitigate threats or impacts are non-existent, vague, or have low effectiveness. Actions are needed to move into positive status and trend.
	Not rated as there is not enough data to produce an assessment.

*Ratings are qualitative, subjective and are designed to provide the reader with a status at a glance.

Table 2. Scheme assessment for the rating of anthropogenic threats and environmental impacts on coastal jaguars from the BPCB and "Cordillera" Chongón-Colonche Range in Ecuador.

much as about the need for actions related to any threat as well as about the current status of any particular issue.

3. Results and discussion

3.1. Field observations: indirect and direct records

The findings of this research highlight the remarkable presence of a very small population of jaguars ($n = 2$) as revealed by the camera traps, field observations (presence of scats, footprints) and forest ranger personal encounter accounts with jaguars around BPCP and surrounding areas.

3.2. Transects surveys

The monitoring along transects provided a preliminary and effective approach to track the existence of jaguars and confirmed the presence of several species of mammals in the BPCP. On transect 3 at the Candil Creek, feline scats were recorded on 22 September 2008 (**Figure 2A**), while on transect 5 another excreta was identified from the same individual on November 26, 2008. Additional inspections conducted at Holcim mining facilities revealed the presence of footprints from a big cat that often travels through the site, but remains distant from the limestone extraction activity and machinery; which is a human-disturbed sector with the presence of an artificial lagoon/pond (**Figure 2B**). For instance, during the first inspection around Holcim (25 September 2008), foot prints (8–9 cm length; **Figure 2C**) and tracks of a larger cat (**Figure 2D**) were observed without being able to confirm the species (i.e. puma or jaguar although cougars have not been recorded for several years at BPCB) because the footprints were slightly deteriorated due to the soil erosion and dryness; however, machinery operators working at this sector confirmed the observation of a jaguar approximately 4 years earlier and their descriptions of the animal observed agree with the characteristics and colours of the jaguar. In the second inspection (15 December 2008), some tracks (footprint length = 8–9 cm; **Figure 2E**) with a footprint inter-distance of 48 cm from the anterior to posterior footprints (**Figure 2F**) were again observed and likely to be left by a jaguar the day following a light rain on December 13. The indirect records on these locations are fully consistent with indirect observations previously collected in the same sites and contiguous sectors by forest rangers in recent years [16], suggesting that this specific individual uses these trails for daily movements to enter the BPCB.

3.3. Photos from camera traps

From June 2008 to February 2009, the deployment of camera traps generated positive outcomes for the photo-records, allowing to mark recapture at least one jaguar (**Figure 3A–D**). A trap camera captured and confirmed the images of the same jaguar in four occasions in 2008 (i.e. September 12, September 16, November 10 and November 11; as shown in **Figure 3**). The trail used by the photo-identified jaguar coincides with the indirect records reported by rangers and with the excreta collected on 22 September 2008 (**Figure 2A**) in a ravine ("Quebrada Candil"),

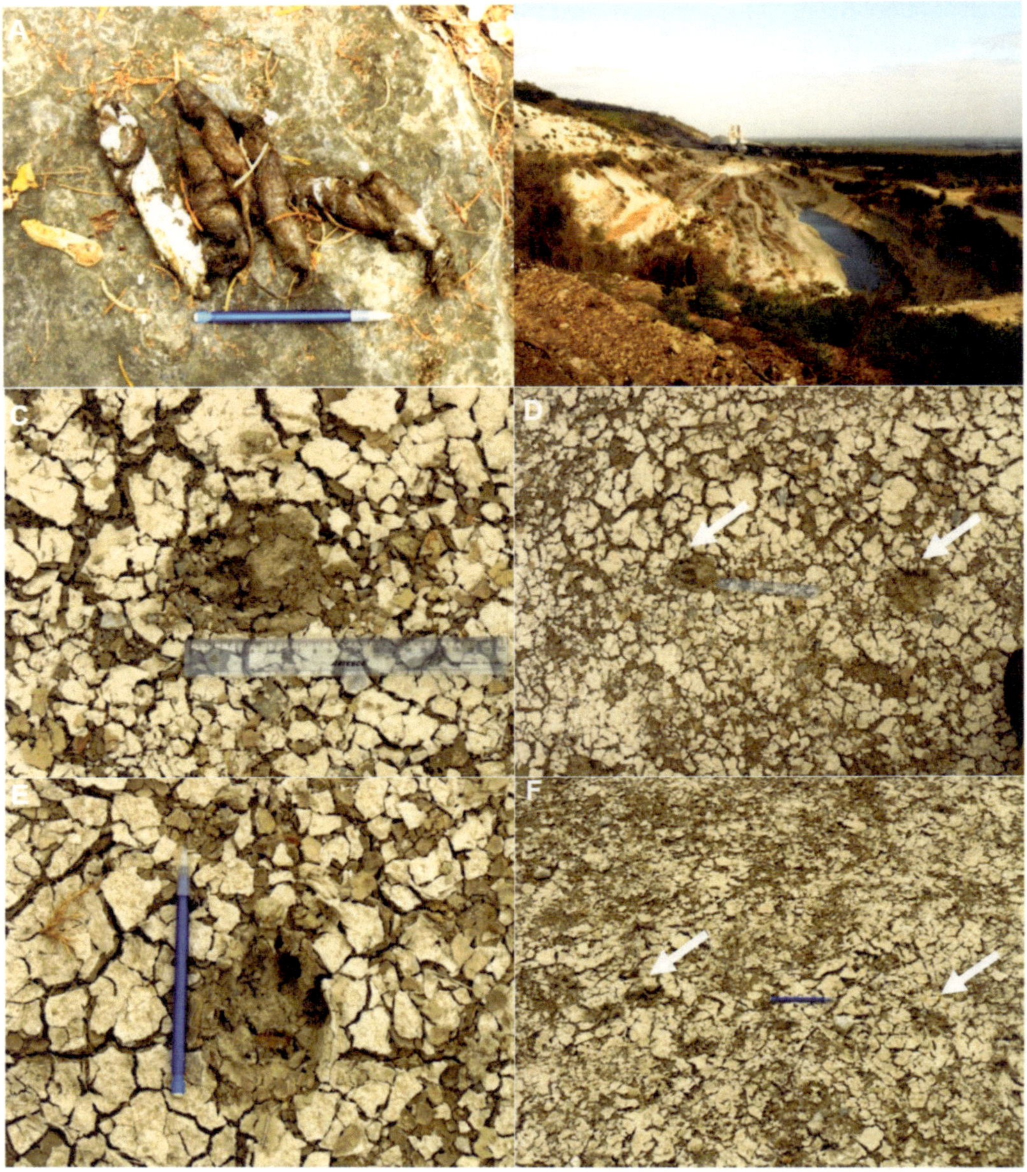

Figure 2. Photos of scats and footprints of a jaguar roaming in and around trails of BPCB. (A) Scats found on transect 3 at the Candil Creek; (B) a photo of the view of the monitored areas at the limestone quarry mine (Holcim installations); (C & D) footprint (8–9 cm) and track (inter-footprint distance = 35–38 cm) recorded on a trail nearby Holcim facilities on 25 September 2008); and, (E & F) additional jaguar's tracks (inter-footprint distance = 43 cm) and footprints (8–9 cm) found in the same site on 15 December 2008. Photos adapted from Ref. [16].

as well as the footprints found around the quarry installations on 25 September 2008 (**Figure 2C** and **D**). In 2011, new images of the same individuals were captured by the camera on April 1, July 12 and July 25 following further field monitoring of the jaguar population at BPCB (**Figure 3E** and **F**).

Figure 3. Camera trap photos of a jaguar mark-captured and recaptured in the study area from 2008 to 2011. The same individual was observed in the following dates in two different sites: (A) 9 September 2008; (B) 16 September 2008; (C) 10 November 2011; (D) 11 November 2011; (E) 01 April 2011; (F) 12 July 2011; and, (G) 25 July 2011. The red and black arrows in photos A-D indicate number and shape of black spots to identify the jaguar. Photos of jaguar captured from 9 September 2008 to 11 November 2008 were adapted from Ref. [16].

3.4. Abundance estimates

Based on the overall field observations, it is estimated that this individual uses 60% of BPCB (approximately 3600 ha) as part of its territory [16]. Conversely, the occurrence of at least one jaguar in the total area of BPCB (i.e. 6000 ha or 60 km^2) would imply a density of 1 jaguar per 60 km^2. This abundance is very similar to the minimum abundance found for this species (1 jaguar per 64 km^2) in the Pantanal, Brazil [33]. A population density of 2.63 jaguars/100 km^2 was estimated at the Cotacachi-Cayapas Ecological Reserve (2500 km^2, including buffer zone) in northwest Ecuador [17], while a density of 2.25 individuals per 10,000 ha (100 km^2) was estimated in Costa Rica [34]. This could demonstrate that the jaguar identified in this study could also include areas within the protected area that are outside the BPCB, and that this area is likely to be a territorial encounter for two individuals, coinciding with the record of a major fighting event or courtship involving two jaguars observed in a ravine near the Jaguar guard station by a forest ranger in 2006 (see **Table 4**).

The field studies on jaguar abundances conducted in Central and South America indicate absolute abundances ranging from 1 to 9 jaguars per 100 km^2 [17, 32, 35–38]. Assuming that the lower bound of this abundance range (i.e. 1 jaguar/100 km^2) can be extrapolated to the total area of the Chongón Colonche Protected Forest (440–700 km^2), it is estimated that a plausible abundance of at least 4–7 jaguars would be inhabiting this area. In the Cotacachi-Cayapas Reserve, it is likely that a small population of at least 20–30 jaguars remain there [13]. The projected jaguar conservation units in western Ecuador account for 10% (i.e. 8700 km^2) of the original distribution and home range for jaguars, as reported by Espinosa et al. [13], with a potential overall population of less than 250 individuals [15]. However, caution should be take into account to interpret these estimations as the forest and basin conditions and environmental factors of these regions differ.

3.5. Associated mammalian fauna

Both direct and indirect records observed along transects revealed the presence of several mammalian species part of the tropical fauna found in BPCP (**Table 3**). As shown in **Figure 4**, these observations are further corroborated based on images of mammals captured from camera traps deployed along transects over the study period at BPCB, where mammals such as white-tailed deer, *Odocoileus peruvianus* (**Figure 4A**); crab-eating raccoon, *Procyon cancrivorus* (**Figure 4B**); ocelot, *L. pardalis* (**Figure 4C**); jaguarondi, *Puma yagouaroundi* (**Figure 4D**); Northern tamandua, *T. mexicana* (**Figure 4E**); Central America agouti, *Dasyprocta punctata* (**Figure 4F**); tayra, *Eira barbara* (**Figure 4G**); forest rabbit or tapeti, *Sylvilagus brasiliensis* (**Figure 4H**); and, white-nosed coati, *Nasua narica* (**Figure 4I**) were photo-identified [16]. Relative to the direct observations and indirect records, the camera trap captured images of most terrestrial mammals, except for two common species of monkey, including the Ecuadorian mantled howler monkey (*Alouatta palliata aequatorialis*) and Ecuadorian white-fronted capuchin (*Cebus aequatorialis*), which are arboreal mammals. Most of these mammalian species, including white-tailed deer, collared peccary, agouti and tapetis, can well serve as prey given the broad spectrum diet and opportunistic predatory behaviour of this species [1, 3, 39, 40]; thus, these mammals reflects the available prey community in the region and can be part of the jaguar's prey to sustain a local population in the BPCB. For instance, as part of a Conservation International's

Common name	Latin name	Direct	Indirect	Camera trap
White-tailed deer	*Odocoileus peruvianus*	P	P	P
Collared peccary	*Pecari tajacu*	P	P	A*
Crab-eating raccoon	*Procyon cancrivorus*	P	P	P
Central America agouti	*Dasyprocta punctata*	P	A	P
Tapeti/forest rabbit	*Sylvilagus brasiliensis*	P	P	P
Mantled Howler monkey	*Alouatta palliata aequatorialis*	P	A	A
White-fronted capuchin	*Cebus aequatorialis*	P	A	A
White-nosed coati	*Nasua narica*	P	A	P
Northern tamandua	*Tamandua mexicana*	P	A	P
Coastal jaguar	*Panthera onca centralis*	A	P	P
Ocelot	*Leopardus pardalis*	A	P	P
Jaguarondi	*Puma yagouaroundi*	A	A	P
Tayra	*Eira barbara*	A	A	P

Adapted from Ref. [16]. P = presence; A = absence.
*A camera trap recently installed at BPCB captured photos of collared peccaries on 24 December 2016 and 6 January 2017.

Table 3. Type of field records and observations for jaguars and associated mammals from 2008 to 2009.

rapid assessment program (RAP) carried out in BPCB, L. Emmons reported the presence of large quantities of bones of deer (i.e. both white-tailed, **Figure 4A**, and red brocket deer, *Mazama americana*, are found in BPCB), indicating the presence of jaguars (E. Horstman, *personal communication*, 2017). Although there are puma (*P. concolor*) in the BPCB, there have been no reported sightings in several years (E. Horstman, *personal communication*, 2017), which may explain the lack of this species in photo-records captured by camera traps.

3.6. Accounts of forest ranger reports and local community testimonies

The recent history of accounts and encounters supporting the existence of jaguars at BPCP started with one of the first confirmed sightings dating back to the early 1990s when the late E. Aspiazu Estrada, then President of the Guayaquil Chapter of Fundación Natura (one of Ecuador's first major environmental organization), observed a jaguar crossing a road below a sightseeing site (known as "Mirador de los Monos") in BPCB (E. Horstman, *personal communication*, 2017). In 1990, the third and fourth authors (E.H. and S.C.) set up a field research and camped nearby this site, but while they did not observed jaguars, they did find large foot prints in the access road near the campsite. In the following years, more tracks on dirt roads, scrapes on trees and occasional scat with soil heaped on the deposit in cat-like fashion have periodically been found (E. Horstman, *personal observations*, 1990s). Several plaster casts have been made of tracks (footprints) and a scat sample was collected. These samples were sent to R. Williams (former country director of the Wildlife Conservation Society), who confirmed that both tracks and the scat were probably jaguar.

Figure 4. Camera trap photos of tropical mammals recorded in the study area from June 2008 to February 2009. (A) White-tailed deer (*O. peruanus*);(B) crab-eating raccoon (*P. cancrivorus*); (C) ocelot (*L. pardalis*); (D) jaguarondi (*P. yagouaroundi*); (E) Northern tamandua (*T. mexicana*); (F) Central America agouti (*D. punctata*); (G) Tayra (*E. barbara*); (H) forest rabbit/ tapeti (*S. brasiliensis*); and, (I) white-nosed coati (*N. narica*). Additional deployment of a camera trap in 2016 confirmed the presence of collared peccaries (*P. tajacu*) in BPCB (J). The photo set containing Figures A-I were adapted from Ref. [16].

The data resulting from the accounts by forest rangers and interviewed park guards are reported in **Table 4**. 100% of the interviewed rangers had observed jaguars or/and experienced at least one encounter with jaguars. One of the captivating forest ranger observations is that of the jaguar attack on a longhorn steer ("criollo" breed) that wandered into an area alone near the Papagayo guard station in the BPCB buffer zone in 1996 (**Table 4**). The steer was attacked by a large cat (jaguar) that left deep scratch marks along the side of the neck and flanks. The jaguar had apparently attacked the steer from behind trying to unsuccessfully kill it by biting behind the head.

In 2006, a park guard reported hearing and watching two large jaguars snarling and either fighting or perhaps in courtship in a ravine near the Jaguar guard station. The same park guard that witnessed the attack on the longhorn steer also had a compelling close encounter face to face with a jaguar in 2007 while riding a horse on patrol on the trail linking the Pigio and Cusumbo guard stations. As the ranger came around a bend in the trail, the horse was startled by a jaguar standing in the middle of the trail and reared and tried to buck off the ranger as it ran away. Similarly, a park guard had a close encounter with a jaguar at night at an encampment where forest restoration was being carried out. The ranger went out and

Number	Interviewed people	Site	Account/testimony/comment
1	Forest ranger/park guard: A.M	Papagayo guard station; BPCB buffer zone	In 1996, a longhorn steer was attacked by a jaguar.
2	Forest ranger/park guard: P.O.	Ravine near the Jaguar guard Station	In 2006, two jaguars were heard and seen snarling and fighting or in courtship.
3	Forest ranger/park guard: A.M.	Trail linking the Pigio and Cusumbo Top guard stations	2007, a close encounter with a jaguar on a trail while ranger was horseback riding.
4	Forest ranger/park guard: F. M.	Forest restoration site	Ranger had a jaguar encounter close to encampment site at night
5	Chief forest ranger/park guard: P.Y.	N/A	Anecdotic information and stories about the confirmed presence of jaguar in BPCB.
6	Property owner and workers	Land within north side of BPCB	A female jaguar and a cub were spotted near an artificial water hole.
7	Caretakers of ranch (Hacienda Molino)	Ranch (Hacienda Molino); BPCB buffer zone	Regular sightings of jaguars are made associated with loss of domestic dogs to big cats.
8	Farmer/rancher	Cerro Azul (Monte Sinai)	A pregnant (female) jaguar had been shot and killed. The farmer claimed that he had been supposedly attacked by the jaguar and killed in self-defense. However, this is questionable as there are no reported attacks by jaguars in and around BPCB.*

**Note:* This report was originally provided by a past director of the Wildlife Management Program of the former governmental organization INEFAN (Instituto Ecuatoriano Forestal y de Areas Naturales), and corroborated later on by a former Fundación Pro-Bosque employee (N. Zambrano).

Table 4. Qualitative data of interviews and accounts from forest rangers and rural communities around BPCB, BPCB buffer zone and surrounding areas.

illuminated a jaguar with his flashlight and he struck a nearby tree with his machete several times, but the jaguar was unperturbed and slowly turned around and walked away.

Recurrent sightings of jaguars are also made in the BPCB buffer zone, including a nearby ranch "Hacienda El Molino," where caretakers have reported the periodic loss of domestic dogs to big cats. Of particular importance are two accounts by a farmer and property owner with his workers from the rural community in and around BPCB as they reported female jaguars either pregnant or with a cub, respectively (**Table 4**). These reports suggest that reproduction for this species may be feasible within a breeding ground in the BPCB.

In general, throughout the development of surveys and exchange of information to search for evidences of the species presence and encounters with members of communities, most of the interviewed persons were likely to confirm the existence of the jaguar, commonly called "tiger" because of their black spots and the similarity that they have noticed by the old films or movies, showing its ferocity at attacking people [16]. Thus, a general "fear factor" has been attributed to this species by the rural community without knowing it.

Meanwhile, the information collected from our surveys with the rangers and rural community partially diverge from the opinion from local residents of Machalilla National Park, where there are mixed perceptions about the presence of jaguar, i.e. some people argued that the jaguars disappeared 20 years ago, while others suggested they still inhabit isolated areas of the park [18].

3.7. Anthropogenic threats

A preliminary assessment of anthropogenic impacts and environmental stressors on coastal jaguars in Ecuador's mainland coast, focused on the BPCB and "Cordillera" Chongón Colonche Range, is shown in **Table 5**. This evaluation scheme reflects a qualitative effort to further advance our understanding of past, present and future impacts and threats jeopardizing the long-term survival of jaguars. The hunting of jaguar for the skin trade and habitat loss by population growth and urbanization, deforestation and mining have been among the most critical impacts negatively affecting jaguar populations in Ecuador [8, 14–16]. Because of the jaguar's large size (i.e. the largest among the New World felids) and the beauty of its skin, this species was and is still one of the most persecuted and hunted mammals by indigenous people, military, farmers, settlers or poachers in the Ecuadorian territory [8, 13, 15]. Conversely, the jaguars are persecuted in many livestock and farming areas as they can sometimes kill cattle and domestic animals or for the fear of the local people to be attacked. At the BPCB and "Cordillera" Chongón Colonche hills, there are only historical anecdotes among the rural community that the species is a threat to this activity, which must be further corroborated. However, the reported isolated jaguar attack on a longhorn steer in 2006 and the presumably attack to a farmer at BPCB (**Table 4**) warrant the need of further research.

Western and Tumbesian forests of Ecuador have been logged and deforested by timber exploitation, farming and ranching, with less than 5% of the original forest coverage remaining by the 1990s [27, 41, 42]. As a result, the extent and degree of deforestation and forest land converted to farms and livestock have proportionally reduced habitat suitability for coastal jaguars in recent times (**Table 5**). Efforts to reforest and rehabilitate areas previously cleared out are underway through forest restoration program at BPCB. While mining for extraction of limestone from

Threats/impacts	Past	Current	Future
Habitat loss and deforestation	(1)	(2)	*(biological corridors implementation is uncertain)
Hunting and illegal traffic	(1)	(3)	*(it could follow the same trend as in Current assessment)
Jaguar's prey poaching and hunting	(1)	(2)	*(it could follow the same trend as in Current assessment)
Mining	(1)	(2)	(4)
Livestock	(1)	(4)	(4)
Pollution of River Basins	(1)	(3)	*(implementation of pollution control and command regulations is uncertain)
Introduced species	(1)	(2)	*(some invasive plant species (e.g., *Panicum maximum*) are controlled within BPCB, but it is unknown whether eradication would take place outside of BPCB and buffer zone)
Diseases	Lack of data	Lack of data	Lack of data
Climate change	(1)	(3)	(3)**

See **Table 2** for definition of colour and rating meanings, as aforementioned.

Remarks:

(Red: 1): Negative impacts and/or threats were or are high risk and have resulted in an endangered status of the species.

(Red: 2): Improvements are uncertain, minor, or slow.

(Red: 3): Actions to address or mitigate threats or impacts are non-existent, vague, or have low effectiveness. Actions are needed to move into positive status and trend. (Red: 3**): The looming desertification in areas with prolonged droughts may cause severe impacts on jaguar's habitat and prey availability in the future. Actions are needed to move into positive status and trend.

(Blue: 1): The status of the forest ecosystem was healthy according to available data (see references [27, 41]) and best available kneeled and judgment from researchers and natural resource managers.

(Blue: 4): Actions to address or mitigate impacts are or will be underway and are likely to be potentially effective (i.e., abandonment and restoration plan following closure of quarry mining and presence of livestock is low).

*Status, trend, data, and/or actions provide contradictory or inconclusive information. Actions are needed to avoid negative status and trend.

Table 5. Snapshot assessment of anthropogenic threats for coastal jaguars in Ecuador's mainland coast focused on the BPCB and "Cordillera" Chongon-Colonche Range.

the BPCB has evidently fragmented and disturbed habitat for jaguars residing in the forests the BPCB and "Cordillera" Chongón Colonche range, the species still persists there although in very low numbers, as documented in this study and elsewhere [16]. It is expected that ecological restoration under an abandonment and environmental management plan will be implemented following the closure of the quarry in the final phase of the project (**Table 5**).

3.8. Current legislation and management actions to conserve jaguars in Ecuador

In Ecuador, the environmental and wildlife legislation to manage and protect jaguars emerged in 1970 when a government decree considered the jaguar as a species of limited hunting under certain regulations of the Ecuadorian state (Official Register No. 818; November 20, 1970), but jaguar hunting was fully banned later on by the Ecuadorian government when the state ratified the Convention on International Trade in Endangered Species of Wild Fauna and Flora (CITES) (Official Register No. 746; February 20, 1975) [15]. Recently, the Ministry of the Environment, through Resolution No. 105, promulgated a law prohibiting hunting of jaguars in the Ecuadorian territory, indefinitely (Official Register No. 5; January 28, 2000). The jaguar's original distribution also overlaps important natural areas protected by the National System of Protected Areas of Ecuador's Minister of Environment. Internationally, The IUCN considers the jaguar to be a Near-Threatened species (NT) [5], while in Ecuador the population of coastal jaguar is considered as critically endangered (CR) [15]. As a signatory country of CITES, the jaguar is listed in Appendix I (https://cites.org/eng/gallery/species/mammal/jaguar.html), prohibiting any commercial activity, including live animals, dead or any of its parts or organs (i.e., jaguar paws, teeth and other products) in Ecuador.

As for our study area, Cerro Blanco Protected Forest (BPCB) was created and managed by the Pro-Bosque Foundation, a private and non-profit organization established under Ministerial Agreement by the Ministry of Agriculture and Livestock, on November 9 (1992), to protect and restore the forest and its river basin. Therefore, by protecting the forest remnants and basin of BPCB, suitable habitat for jaguars and their prey can be conserved in the long-term. Similarly, the "Cordillera" Chongón-Colonche Mountain Range was declared Protected Forest in 1994 with the aim of achieving conservation of the pre-mountain Humid Forest and Tropical Dry Forest [43]. Since 1998, forest plantations have been fostered under different systems in the protected forest's buffer zone, as a management strategy to counteract the extraction of wood/timber [43].

In addition, the jaguar is included under the category of threatened species by the Environmental Management Ordinance of the Provincial Council of Guayas Province, involving the protection of threatened (or endangered) species with the aim of recovering its population stability. Likewise, the Municipality of Guayaquil passed the Environmental Policies of the Municipality of Guayaquil, in which the jaguar is considered a native species of tropical dry forests and thus included in policy 4.5, involving the recovery of this species.

3.9. Conservation implications

The indirect records of jaguar, ranger encounters and the photo-identified individual inhabiting the protected tropical forest (i.e. BPCB) and surrounding areas of the "Cordillera" Chongón-Colonche Mountain Range provide strong evidence of jaguar in our study region. The individuals found here face different degrees of anthropogenic impacts and fragmentation [16]. Moreover, in the face of lack of additional information available to manage this species, the findings and theoretical estimates of plausible abundance provided here can be used as a guide and premise to support decision makings and policy aimed to foster *in situ* conservation and proactive management actions for the species, following the precautionary principle. Because jaguars still remaining in these threatened regions show resilience and adaptation to human perturbations, conservation initiatives and adaptive management actions should consider the region as high priority jaguar conservation unit to preserve one of the last jaguar populations of the Ecuadorian coast. The inclusion of private landowner and ranch owners in jaguar conservation programs with proactive conservation actions for jaguars' habitat and potential corridors for their dispersal within critical habitats may promote long-term population viability by functionally adding private land and ranches to reserve size within jaguar conservation units [44–47]. To achieve this community-based conservation level, proactive relationships between ranchers and authorities must be established to maintain effective agreements [47]. Moreover, because of the small population size, isolation, defective protections and escalating human population, future conservation efforts by Red List assessments' assessors can be prioritized for the most threatened subpopulations of jaguars as an imperative subject of urgency [6]. Thus, contrasting to the "low priority conservation unit" attributed to the habitat of coastal jaguars in the Cotacachi-Cayapas, Cordillera Chongón-Colonche and BPCB [13], we strongly argue that these regions are one of the most important areas and critical habitats of most urgent conservation concern to protect the long-term survival of coastal jaguar subpopulations.

Acknowledgements

We specially thanks the late Eduardo Aspiazu Estrada (Fundación Natura's Guayaquil Chapter) for his deep commitment to conserve BPCB and providing the first sightings of jaguar, as well as the following BPCB parks guards/forest rangers, Armando Manzaba, Felipe Martinez, Policarpio Ortiz and Don Pefecto Yagual and the rural community around BPCB for reporting and sharing jaguar observations and testimonies. We also thank Soraya Delgado, Jacqueline Sócola and Nelson Zambrano for field work assistance. We are thankful for the funding for field research and jaguar monitoring was provided by Conservation International (Ecuador), La Cemento Nacional (Holcim) and Fundación Pro-Bosque. Bob Henderson provided funding for the field work in 2008. We are in debt with Fabian Viteri for his remarkable technical assistance to develop the map for jaguar sightings and records in the BPCB and surrounding areas. J.J. Alava acknowledges the Simon Fraser University (SFU) Open Access Fund, specially Rebecca Dowson, to make possible the permanent unrestricted online access and cover the processing charges of this contribution.

Author details

Miguel Saavedra Mendoza[1], Paúl Cun[2], Eric Horstman[2], Sonia Carabajo[2] and Juan José Alava[3,4,5]*

*Address all correspondence to: jalavasa@sfu.ca

1 Urbanización Paraíso del Río, Guayaquil, Ecuador

2 Fundación Pro-Bosque, Bosque Protector Cerro Blanco, Guayaquil, Ecuador

3 School of Resource and Environmental Management, Simon Fraser University, Burnaby, British Columbia, Canada

4 Institute for the Oceans and Fisheries, University of British Columbia, Vancouver, British Columbia, Canada

5 Coastal Ocean Research Institute, Vancouver Aquarium Marine Science Centre, Vancouver, British Columbia, Canada

References

[1] Seymour KL. *Panthera onca*. Mammalian Species. 1989;**340**:1-9

[2] Nowell K, Jackson P, editors. Wild Cats. Status Survey and Conservation Action Plan. Gland, Switzerland and Cambridge, UK: IUCN/SSC Cat Specialist Group; 1996. p. 406. http://www.carnivoreconservation.org/files/actionplans/wildcats.pdf

[3] Emmons L. Neotropical Rainforest Mammals: A Field Guide. 2nd ed. Chicago, USA: University of Chicago; 1997. p. 396

[4] Sanderson E, Redford KH, Chetkiewicz C, Medellin RA, Rabinowitz A, Robinson JG, Taber A. Planning to save a species: The jaguar as a model. Conservation Biology 2002;**16**: 58-72. DOI: 10.1046/j.1523-1739.2002.00352.x

[5] Caso A, López-González C, Payan E, Eizirik E, de Oliveira T, Leite-Pitman R, Kelly M, Valderrama C. *Panthera onca*. The IUCN Red List of Threatened Species, 2008. Available from: http://dx.doi.org/10.2305/IUCN.UK.2008.RLTS.T15953A5327466.en. [Accessed: April 6, 2017]

[6] De la Torre JA, González-Maya JF, Zarza H, Ceballos G, Medellín RA. The jaguar's spots are darker than they appear: Assessing the global conservation status of the jaguar *Panthera onca*. Oryx. 2017;1-6. DOI: https://doi.org/10.1017/S0030605316001046

[7] Wildlife Conservation Society. All about Jaguars: Threats. New York, USA: WCS; 2008

[8] MAE and WCS. Plan de acción para la conservación del Jaguar en el Ecuador. Quito, Ecuador: Ministerio del Ambiente del Ecuador, Wildlife Conservation Society, Liz Clairborne and Art Ortenberg Foundation, and Wild4Ever; 2014. p. 37. http://www.wild4ever.com/wp-content/uploads/2012/05/Ecuador-National-Jaguar-Plan.pdf

[9] Tirira D. Mamíferos del Ecuador. Quito, Ecuador: Pontificia Universidad Católica del Ecuador, and SIMBIOE; 1999. p. 392

[10] Albuja L, Arcos R. Lista de Mamíferos actuales del Ecuador. Revista Politécnica (Serie Biología 7) 2007;**27**(4):7-33

[11] Tirira DG. Guía de campo de los Mamíferos del Ecuador: Publicación especialsobre los mamíferos del Ecuador 6. Quito, Ecuador: Ediciones Murciélago Blanco; 2007. p. 576

[12] Albuja L. Lista de mamíferos actuales del Ecuador. Quito: Instituto de Ciencias Biológicas, Escuela Politécnica Nacional. 2011:p. 27. http://bibdigital.epn.edu.ec/handle/15000/3843

[13] Espinoza S, Albuja L, Zapata-Rios G, Araguillin E, Utreras V, Noss A. Análisis del estado de conservación del jaguar en Ecuador. In: Medellín RA, de la Torre JA, Zarza H, Chávez C, Ceballos G, editors. El Jaguar en el Siglo XXI: la perspectiva continental. Mexico City, Mexico: Fondo de Cultura Económica, Instituto de Ecología, Universidad Nacional Autónoma de México; 2016. pp. 319-338

[14] Saavedra M. Determinación del uso de hábitat del Jaguar (*Panthera onca*) en el Bosque Protector Cerro Blanco, Cordillera Chongón Colonche. Informe Técnico de Actividades. Guayaquil: Fundación Pro-Bosque. Junio del; 2009

[15] Espinosa S, Zapata G, Saavedra M, Alava JJ, Tirira D. Jaguar (*Panthera onca centralis*). In: DG Tirira, editor. Libro Rojo de los Mamíferos del Ecuador, 2da edición. Publicación especial sobre los mamíferos del Ecuador 8. Quito, Ecuador: Fundación de Mamíferos y Conservación, Pontificia Universidad Católica del Ecuador y Ministerio del Ambiente del Ecuador; 2011. pp. 94-95

[16] Saavedra-Mendoza M, Cun P. Estado actual de conservación, primer registro fotográfico y estimación poblacional del Jaguar de la Costa (*Panthera onca centralis*) y registros de fauna acompañante de mamíferos medianos y mayores en el Bosque Protector Cerro Blanco de la Cordillera Chongón Colonche. Guayaquil—Ecuador. *Revista Científica Yachana*. 2013;**2**(2):127-138

[17] Zapata-Ríos G, Araguillin E. Estado de conservación del jaguar y el pécari de labio blanco en el Ecuador occidental. Revista de Biodiversidad Neotropical. 2013;**3**(1):21-29. DOI: 10.18636/bioneotropical.v3i1.117

[18] Cervera L, Lizcano DJ, Parés-Jiménez V, Espinoza S, Poaquiza D, de la Montaña E, Griffith DM. A camera trap assessment of terrestrial mammals in Machalilla National Park, western Ecuador. Check List. 2016;**12**(2):1868. DOI: http://dx.doi.org/10.15560/12.2.1868

[19] Albuja VL. Diagnostico faunístico para la actualización del plan de manejo del Parque Nacional Machalilla. Quito: Instituto Ecuatoriano Forestal de Áreas Naturales y Vida Silvestre; 1997: p. 62

[20] Araguillin E, Zapata-Ríos G. Muestreo con trampas fotográficas de mamíferos medianos y grandes en el refugio de vida silvestre El Pambilar y dos localidades de la reserva ecológica Cotacachi Cayapas, informe técnico, Quito: Wildlife Conservation Society—Programa Ecuador; 2011

[21] Porras PI. Nuestro ayer: Manual de arqueología ecuatoriana. Quito, Ecuador: Centro de Investigaciones Arqueológicas del Ecuador, Pontificia Universidad Católica del Ecuador (PUCE); 1987

[22] Peregrine PN, Ember M. Encyclopedia of Prehistory. Vol. 7: South America. New York: Kluwer Academic Plenum Publishers; 2001. p. 194

[23] Myers N, Mittermeier RA, Mittermeier CG, Da Fonseca GA, Kent J. Biodiversity hotspots for conservation priorities. Nature. 2000;**403**(6772):853-858. DOI: 10.1038/35002501

[24] Tirira DG, editor. Libro Rojo de los mamíferos del Ecuador. 2nd ed. Publicación especial sobre los mamíferos del Ecuador 8. Quito, Ecuador: Fundación Mamíferos y Conservación, Pontificia Universidad Católica del Ecuador y Ministerio del Ambiente del Ecuador; 2011: p. 398. www.librorojo.mamiferosdelecuador.com

[25] Horstman E. Plan de Manejo Bosque Protector Cerro Blanco. Guayaquil: La Cemento Nacional, Fundación ProBosque, Bosque Protector Cerro Blanco e INEFAN. 30 de abril de; 1998

[26] Horstman E, Carabajo S. Estrategia para el establecimiento de un corredor biológico para la conservación del Bosque Protector Cerro Blanco con énfasis en el Guacamayo

Verde Mayor *Ara ambigua guayaquilensis* [thesis]. Guayaquil: Universidad Agraria del Ecuador Sistema de Post Grado SIPUAE; 2005

[27] Best BJ, Kessler M. Biodiversity and conservation in Tumbesian Ecuador and Peru. Cambridge, United Kingdom: BirdLife International. 1995: p. 218

[28] BirdLife International. Important bird area factsheet: Bosque Protector Chongón-Colonche, Ecuador: BirdLife International; 2009. Available from: http://www.birdlife.org [Accessed: 30-06-2009]

[29] Cuesta F, Peralvo M, Sánchez D. Métodos para investigar la disponibilidad del hábitat del oso andino: el caso de la cuenca del río Oyacachi, Ecuador. Serie Bioreserva del Cóndor no. 1. Quito, Ecuador: Ecociencia y Proyecto Biorreserva del Cóndor; 2001: p. 67

[30] Aranda M. Huellas y otros rastros de los mamíferos grandes y medianos de México. Xalapa, México: Instituto de Ecología, A.C. y la Comisión Nacional para el conocimiento y Uso de la Biodiversidad; 2000. p. 212

[31] Silver S. Estimando la abundancia de jaguares mediante trampas cámara. Programa para la conservación del Jaguar. New York, USA: Wildlife Conservation Society; Diciembre del. 2004. p. 27

[32] Silver S, Ostro L, Marsh L, Maffei L, Noss A, Kelly M, Wallace R, Gómez H, Ayala G. The use of camera traps for estimating jaguar *Panthera onca* abundance and density using capture/recapture analysis. Oryx. 2004;**38**:148-154. DOI: https://doi.org/10.1017/S0030605304000286

[33] Quigley HB, Crawshaw PG. A conservation plan for the jaguar *Panthera onca* in the Pantanal region of Brazil. Biological Conservation. 1992;**61**:149-157. DOI: https://doi.org/10.1016/0006-3207(92)91111-5

[34] González-Maya JF. Densidad, uso de hábitat y presas del jaguar (*Panthera onca*) y el conflicto con humanos en la región de Talamanca, Costa Rica [thesis]. Turrialba, Costa Rica: Programa de Educación para el Desarrollo y la Conservación, Escuela de Postgrado, Centro Agronómico Tropical de Investigación y Enseñanza (CATIE); 2007

[35] Wallace RB, Gómez H, Ayala G, Espinoza F. Camera trapping for jaguar (*Panthera onca*) in the Tuichi Valley, Bolivia. Mastozoologıa Neotropical. 2003;**10**:133-139. http://www.osi-perception.org/IMG/pdf/Camera_trapping_for_jaguar.pdf

[36] Maffei L, Cuéllar E, Noss A. One thousand jaguars (*Panthera onca*) in Bolivia's Chaco? Camera trapping in the Kaa-Iya National Park. Journal of Zoology. 2004;**262**:295-304. DOI: 10.1017/S0952836903004655

[37] Soisalo MK, Cavalcanti SMC. Estimating the density of a jaguar population in the Brazilian Pantanal using camera-traps and capture-recapture sampling in combination with GPS radio-telemetry. Biological Conservation. 2006;**129**:487-496. DOI: http://doi.org/10.1016/j.biocon.2005.11.023

[38] Salom-Perez R, Carillo E, Saenz JC, Mora JM. Critical condition of the jaguar *Panthera onca* population in Corcovado National Park, Costa Rica. Oryx. 2007;**41**:51-56. DOI: https://doi.org/10.1017/S0030605307001615

[39] Rabinowitz AR, Nottingham BG Jr. Ecology and behaviour of the jaguar *(Panthera onca)* in Belize, Central America. Journal of Zoology. 1986;**210**:149-159. DOI: 10.1111/j.1469-7998.1986.tb03627.x

[40] Sunquist M, Sunquist F. Wild Cats of the World. Chicago, IL: University of Chicago Press; 2002: p. 462

[41] Dodson CH, Gentry AH. Biological extinction in western Ecuador. Annals of the Missouri Botanical Garden. 1991;**78**(2):273-295

[42] Sierra R. Vegetación remanente del Ecuador continental. Quito: Ministerio de Medio Ambiente/Proyecto Instituto Ecuatoriano Forestal y de Areas Naturales y Vida Silvestre, Fondo para el Medio Ambiente Mundial-Banco Internacional de Reconstrucción y Fomento, Ecociencia, Wildlife Conservation Society; 1999

[43] Cañadas A, Roca-Cedeño A. Can the reforestation projects stop the extraction of wood from the protected forest Chongón-Colonche? In: Becker M, Kreye C, Ripken C, Tielker E, editors. Book of Abstracts Annual Conference on Tropical and Subtropical Agricultural and Natural Resource Management (Tropentag 2011); 5-7 October 2011; Bonn, Germany: University of Bonn; 2011. p. 260. http://www.tropentag.de/2011/abstracts/abstracts.php?showtime=0&noID=1&menu=11

[44] Woodroffe R, Ginsberg JR. Edge effects and the extinction of populations inside protected areas. Science. 1998;**280**:2126-2128. DOI: 10.1126/science.280.5372.2126

[45] Cousins JA, Sadler JP, Evans J. Exploring the role of private wildlife ranching as a conservation tool in South Africa: Stakeholder perspectives. Ecology and Society. 2008;**13**(2):43. http://www.ecologyandsociety.org/vol13/iss2/art43/

[46] Rabinowitz A, Zeller KA. A range-wide model of landscape connectivity and conservation for the jaguar, *Panthera onca*. Biological Conservation. 2010;**143**:939-945. DOI: 10.1016/j.biocon.2010.01.002

[47] Gutiérrez-González CE, Gómez-Ramírez MA, López-González CA, Doherty PF Jr. Are private reserves effective for jaguar conservation? PLoS One. 2015;**10**(9):e0137541. DOI: 10.1371/journal.pone.0137541